REPORT
WRITING

for the
Community
Services

REPORT
WRITING

for the
Community
Services

Dianna McAleer

ALGONQUIN COLLEGE

Pearson Canada
Toronto

Library and Archives Canada Cataloguing in Publication

McAleer, Dianna, 1962–

Report writing for community services / Dianna McAleer.
Includes index.

ISBN 978-0-13-206894-9
1. Report writing—Textbooks.   2. Communication in social work—Textbooks.   I. Title.

HV29.7.M22 2010   808'.066361   C2008-904537-8

ISBN-13: 978-0-13-206894-9
ISBN-10: 0-13-206894-X

Vice-President, Editorial Director: Gary Bennett
Editor-in-Chief: Ky Pruesse
Acquisitions Editor: David Le Gallais
Marketing Manager: Loula March
Developmental Editor: Rema Celio
Production Editor: Melissa Hajek
Copy Editor: Deborah Cooper-Bullock
Proofreaders: Marg Bukta, Nancy Carroll
Production Coordinator: Janis Raisen
Composition: Integra
Art Director: Miguel Angel Acevedo
Cover and Interior Designer: Peter Papayanakis
Cover Image: Corbis

11   14

Printed in Canada

For the teachers in my life: my family, my friends,
and my students

BRIEF CONTENTS

CONTENTS

PREFACE

In the community services field, strong communication skills are a must. Regardless of which area of the field you choose to work in, chances are you will be writing logs, letters, reports, and proposals and attending meetings, hearings, and seminars on a regular basis. You will need to be able to write competently, effectively, and professionally, often with little time for editing or second thought. Furthermore, the writing you do will be kept as permanent records and used in the legal system. You will also need to be able to prepare informative and well-organized presentations for a variety of audiences.

This practical, hands-on textbook will give you the skills you need to write memos, letters, email, and reports that meet the professional standards required for documentation and record-keeping in the community services field.

Chapter 7: Presentation Skills will teach you how to plan and deliver organized, informative, and dynamic presentations.

Prepare for your employment search with information and exercises about writing cover letters and resumés. You will learn to prepare for, perform during, and debrief after a community services job interview.

Sample templates have been included for you to read as well as exercises that give you the opportunity to practise and perfect your skills. Each chapter also includes a Sentence Polishing Skill that will help you elevate your writing style to a professional level.

This book concludes with an appendix in which you can find tips on conducting proper research and avoiding plagiarism.

Report Writing for the Community Services was written to help you succeed in the community services field. I hope you find it informative and useful. You've chosen a noble profession; good luck in your college and community services careers.

Supplements

Instructor's Manual

The instructor's manual for this text includes chapter summaries, answers for the Sentence Polishing Skills exercises, and extension activities for the material covered in each chapter. I would appreciate feedback regarding further material you would like to see in future editions of the text and the instructor's manual.

CourseSmart

CourseSmart is a new way for instructors and students to access textbooks online anytime from anywhere. With thousands of titles across hundreds of courses, CourseSmart helps instructors choose the best textbook for their class and give their students a new option for buying the assigned textbook as a lower-cost eTextbook. For more information, visit www.coursesmart.com.

Acknowledgments

I would like to thank the many professionals who have helped shape and polish this book. To the Pearson team—Chris Helsby, Emily Jardeleza, Rema Celio, and Melissa Hajek, thank you for your support and patience. Deborah Cooper-Bullock, Marg Bukta, and Nancy Carroll, thank you for your eagle eyes and your editorial wisdom.

I would also like to thank the professors who reviewed the text. I appreciate your thoughtful suggestions; they have certainly improved the focus and scope of the material. Thank you . . .

Nancy Rishor, Fleming College
Karen Marr, St. Clair College
Ann Mikula, Algonquin College
Chris Harris, Durham College
Louise Bureau, Grant MacEwan College
Carey Vivian, Douglas College

Report Writing Basics

LEARNING OBJECTIVES

After reading this chapter and completing the exercises in it, you will be able to

1. Write objectively using facts to support your observations.

2. Use the active voice.

3. Select an appropriate tone and writing style by editing for jargon, slang, and clichés.

4. Record information in chronological order.

5. Organize information by making appropriate use of headings, subheadings, and white space.

6. Choose appropriate fonts and typefaces for your reports.

Introduction

As a community services worker, you will have hundreds of interactions with clients, colleagues, and other professionals in the field. Some interactions will be brief or casual. For example, as a worker in a halfway house, you might have a 10-minute conversation with a client about her progress in an addictions program. Other interactions will be longer and more formal. For instance, as a worker for a subsidized housing organization, you might conduct a 30-minute assessment interview with a new client. You must record every interaction you have *with* a client in which treatment, behaviour, concerns, successes, requests, or progress is discussed. Similarly, you must record every interaction you have *about* a client with professionals, family members, or other concerned parties.

Would you need to report saying hello to a client as you came on shift? No. If, however, after you say hello to the client, the client reveals a concern she has about her situation, and the two of you spend a minute or two discussing the issue, you would need to record the conversation. You might do this by writing an entry in the client's case file, the facility's daily log, or both.

Reporting client information is an integral part of your job as a community services worker. In this chapter, you will learn the basic principles of reporting information in clear, professional, and informative prose.

SENTENCE POLISHING SKILL: PARALLELISM

Consider this sentence:

> This afternoon, I was able to type three reports, my filing got done, and I also did two intake interviews.

The meaning is clear, but the sentence is awkward.

Here is the same sentence written in parallel structure:

> This afternoon I was able to type three reports, file my papers, and complete two intake interviews.

The revised sentence flows better. The writer has used the same format (present tense verbs) to set up his list of accomplishments.

Making a sentence parallel is straightforward. If you have a list of items in your sentences, make sure that each item has the same grammatical structure.

Examples

You can use -ing verbs:

> I enjoy running, swimming, and boxing.

You can use the infinitive of verbs:

> The youth worker was able to de-escalate the conflict, separate the youth, and restore order in the house.

You can use the same word structure:

> I have three goals in life: to be happy, to be healthy, and to be true to myself.

At the end of this chapter, you will find an exercise to help you practise using parallelism.

Note: Whether you record case notes, intake interviews, or daily logs, your purpose is always the same: to *report* information about your client.

For the sake of simplicity, in this text the term "report" will be used as the generic term for all the writing you do for your clients' files.

Reasons for Writing Reports

Imagine this scenario:

It is a busy day at a drop-in centre for youth. In the centre's lobby, a child and youth worker is informing a new client about the services offered. At the reception desk, a youth is asking a drop-in centre worker for bus tickets so he can go to a job interview. Another youth is sitting on a couch in a corner. She is meeting her probation officer there to discuss her community service orders. A police officer is in another corner talking to a staff member about a youth, Bobby X, who has violated parole conditions and gone missing. The police officer believes the youth has a gun, and since Bobby has used the drop-in centre in the past, the officer believes he may drop in again. Just then, Bobby X comes into the centre. He sees the officer and panics. He pulls out a gun and grabs the youth from the couch, threatening to shoot his hostage if the officer tries anything. The police officer is able to calm Bobby X down and arrests him.

Two months later, the drop-in centre worker and the child and youth worker are both called to court to testify about the event. Neither wrote a report when the event occurred. In fact, the centre worker did not even report the event to her director. The centre worker does not remember who else was in the lobby at the time of the incident. She also does not remember exactly what happened. She has no notes to which to refer. Nor does the child and youth worker remember exactly what happened. She cannot remember which new client she had with her at the time of the incident. She has no reference notes either.

You can imagine the impact the absence of a record would have on the court case and the careers of the workers.

Workers need to write reports for many reasons:

- Workers need to document changes in clients' behaviours, emotional states, and attitudes to monitor, anticipate, and head off problems. These problems could be risk-taking behaviours such as criminal activity or substance abuse, or they could be dangerous behaviours such as self-harming or harming others.
- Your job as a community services worker is to improve the situations of your clients by helping them access the services they need to lead healthy and productive lives. This means developing treatment plans that need to be followed and adjusted when necessary. Treatment plans are effective only when goals and progress are well documented.
- Workers must also protect themselves by documenting their actions and decisions. There will be occasions when you are asked to justify your actions or decisions, and your reports will be invaluable references for you.
- Finally, you must write reports to meet the accreditation, licensing, and legislation standards of your profession.

Writing Style

The key to writing informative, clear, and professional reports is your ability to carefully select words and phrases that are factual and informative.

Here are four guidelines to follow when writing reports:

- Write objectively.
- Write in the active voice.
- Use jargon sparingly.
- Avoid clichés and slang.

Write Objectively

For most of the report writing you will do in the community services field, you will need to write objectively. Objective writing conveys facts and information without emotional wording or an emotional tone. Subjective writing allows the writer's opinions and emotions to be known.

Your workplace writing must be an unbiased narrative of events, conversations, and situations as you experienced them. There will be times when you are frustrated, angry, or disappointed with your client, your colleagues, your organization, or other organizations in the field; however, your emotions must not show in your writing. If your writing does reveal your feelings, your reader may become concerned about your ability to exercise control and use good judgment.

Imagine reading these report excerpts written by various community services workers. As you read them, think about why some of the wording may not be appropriate.

> Today, I met with Samantha. As usual, she was dirty and hadn't bothered to brush her hair. I'm so sick of her attitude.

In this excerpt, the reader knows the writer is frustrated with his client by his use of the words "as usual," "dirty," "bothered," and "I'm so sick." As an advocate for your clients, you must not let their poor decisions personally affect you as it is your job to be their support and constant champions.

Here is an objective version of the same excerpt:

> Today, I met with Samantha. She smelled of body odour; her hair was matted and oily; and her clothes were stained and wrinkled. Since her cleanliness has been an on-going issue (I have noted the same observations in my previous two reports), I started our session with a discussion about the importance of personal hygiene.

In this version, the worker shows that Samantha needs personal hygiene lessons without including his feelings, one way or the other, about the client's unhygienic state. He paints a negative picture of his client's hygiene, but uses no judgmental or subjective language. He states provable, observable facts.

> I would like to know why you thoughtlessly turned down my proposal for funding for a crafts night for our shelter. Do you not understand how important this program is?

In this excerpt, the reader knows the writer is angry at being turned down for a funding request. It is probable that the writer will have to approach that same reader for future funding, so venting her anger in this situation may jeopardize her future success.

Here is an objective version of the same excerpt:

I am disappointed that you were not able to approve funding for my crafts night proposal. This program is very important to our clients, and I would like to know if you would consider discussing the program further. I would appreciate knowing how to improve my proposals so that I will be successful with future funding requests.

In this version, the worker expresses disappointment but does not blame the reader. Instead, she asks for help so that future proposals will be more successful. Asking for help improves her chances at being successful in the future.

I would like to set up a meeting with you, our new staff member Suzanne, and me to discuss her ridiculous behaviour. She has a very unprofessional attitude and dresses like a sex trade worker. Our male clients can't take their eyes off her, and she knows it. She lets them touch and hug her. It's disgusting. They never listen to me when she's around. If something happens to her, it will be her own fault.

In this excerpt, the report writer clearly has a problem with a colleague and wants her boss to know about it. It is a good idea to discuss your valid concerns before they become a big problem. In this case, however, the writer will have to continue working with Suzanne, so she should consider a more diplomatic approach.

Here is an objective version of the same excerpt:

I would like to set up a meeting with you, our new staff member Suzanne, and me to discuss appropriate work ethics. I have witnessed some unprofessional behaviour from Suzanne, and I'm concerned about the impact this will have on the clients and our relationships with them. Suzanne has been allowing some physical contact between her and the clients, and I have seen clients hug her, which, as you know, can be a safety concern.

Here, the worker has expressed the same concerns but has left all personal comments out of her writing. She is writing as a professional concerned about the safety of her colleague and the quality of care given to her clients.

> **Note:** Some reports, such as treatment plans, progress reports, or referral letters, call for your professional recommendations. Your judgment will be based on facts, but it will also include your opinion. In these cases, subjective writing is appropriate as long as you base your opinion on facts and make it clear that you are expressing an opinion, not a fact.

ACTIVITY 1.1 Writing Objectively

Identify and underline the subjective wording in the following report excerpts, and then rewrite the sentences so they are written objectively.

1. Xavier has once again missed his progress meeting. I'm getting to the end of my rope with him. He deserves the consequence I will be giving him. He will not be allowed to

go to the baseball game with the rest of the residents. If that doesn't work, I don't know what I'll do.

_____

_____

_____

_____

_____

2. Sonia's attitude is ridiculous. She refuses to help out with chores because she was spoiled at home. Her parents really messed up with her. I plan to set up a chores schedule for her and to monitor her closely so she sticks to it. That should straighten the little princess out.

_____

_____

_____

_____

3. Jean Luc set the toaster on fire today by trying to heat a hamburger patty in it. The dummy then tried to put the fire out with his bare hands. Can you believe it? The paramedics who bandaged Jean Luc's hands told me Jean Luc had second-degree burns on both palms. What a nut!

_____

_____

_____

_____

_____

Write in the Active Voice

Writing in the *passive* voice omits the doer from the sentence:

The violent youth was successfully restrained.

Writing in the *active* voice identifies the doer(s) of the action in the sentence:

Amelia and Everald successfully restrained the violent youth.

In community services reporting, it is necessary to identify all individuals involved in any action, event, or meeting. Imagine if, two months after the youth noted above was restrained, he suffered a seizure and the hospital staff thought that a blow to the head during the restraint incident caused a blood clot in the youth's brain. The hospital staff needs to question the workers to find out what happened during the restraint. If the workers are identified in a report, they can be questioned and they can assure hospital staff they performed the restraint properly. If the workers are not identified in the report, and no one recalls the incident particulars, the doctors could assume that the facility was at fault, and the facility could face disciplinary action or court charges.

Every time you write reports that describe human involvement, you must be clear about *who* did *what*.

ACTIVITY 1.2   Writing in the Active Voice

Rewrite the following sentences so they are written in the active voice. You will have to add subjects to each sentence, and you may have to change around some wording. The first sentence is done as an example.

1. A cutlery count was performed, and the kitchen was locked for the night.

   I performed a cutlery count and locked the kitchen for the night.

2. When the fire was discovered, 911 was called and the residents were evacuated.

   _____

   _____

3. The youth's mother was called and asked to attend a progress meeting.

   _____

   _____

4. The woman was referred to the Addison Addictions Centre for immediate admission into the residential treatment program.

   _____

   _____

5. The young man's aggressive behaviour was reported to the house director.

   _____

   _____

Use Jargon Sparingly

Jargon can be described as special phrases, short forms, or acronyms used in a given industry. Jargon is useful because it allows people within an industry to save time when speaking and writing. Jargon is only beneficial, however, if everyone involved in the communication process knows the jargon. Imagine that a provincial Member of Parliament received the following request from a local halfway house:

> Dear MP Sarah Patrick,
>
> I am the director of HSC and am writing to ask for your help. Often our clients have trouble completing their CSOs because of the lack of early-morning bus service in our area. Could you help us persuade the bus company to start an early-morning service? The POs with whom we work will support this. I have written to CSC, and they directed me to you.

If you confuse or annoy your reader, you lower your chances of getting what you want.

If you are writing reports that will be read only by colleagues, you can use jargon when appropriate; however, if you are writing reports that will be read by a variety of professionals and non-industry readers, avoid jargon. Explain unique terms, spell out names and titles, and change wording when necessary.

ACTIVITY 1.3 # Discussing Jargon from Your Daily Lives

Get into small groups of three or four. Take turns discussing the jargon you know. See if your classmates know the meaning of this jargon. Use your workplace, sports teams, school, clubs, etc. as resources for jargon.

Avoid Clichés and Slang

Clichés and slang are important additions to a language, and using them makes your speech more colourful and lively. For professional workplace writing, however, you should avoid using both.

Clichés are phrases that people have used over and over again. For example,

- Sick as a dog
- Easy as pie
- Wrong side of the bed
- That takes the cake
- Batting a thousand
- Quiet as a mouse

- Running on empty
- Run like the wind
- Dumb as a post
- Over the top
- Cry like a baby
- Beg and plead

- Moan and groan
- Count your lucky stars
- Beat around the bush
- Better safe than sorry
- Nose to the grindstone
- Happy as a clam

The problem with using clichés is that they make your writing seem unoriginal, and they often do not provide a clear description of what you want to say.

Consider this report excerpt:

My meeting with the board of governors went as smooth as silk. We got on like a house on fire. I thought I'd better be safe than sorry, so I handed each board member a copy of our funding proposal. They were pleased as punch to get that. I think they'll agree to our request for more money for staff training. As soon as they approve the plan, we can run like the wind with it . . .

What is this writer actually saying? Do you feel that this writer is a good writer?

_____

_____

_____

Slang is words or phrases that people have invented either to replace other words or to describe a new phenomenon. Slang can originate in music, films, or fads. Because English is a living language, slang is always developing. It can be so popular that it becomes a part of our standard lexicon. For example, William Shakespeare invented the slang word "fireworks," which is now a recognized word. Conversely, slang can be specific to one region, time period, or age group. This means that people outside that frame of reference will not necessarily understand the slang. Slang also goes out of fashion, so using it can date you.

ACTIVITY 1.4 Using Clichés

Write three or four sentences to describe this morning's activities, from when you got out of bed until you left your home. Use as many clichés in your writing as you can. Take turns with your classmates, and read your work to the class.

Think about the writing you heard your classmates read. Did each person's morning start to sound the same? Did you get a clear picture of exactly how all your classmates spent their mornings? What does "getting up on the wrong side of the bed" mean, anyway?

ACTIVITY 1.5 Using Slang

Take five minutes to write down a list of slang words. Get into a group of four, and take turns sharing your words with the words of others in your group. Have one group member write down the slang that was not understood by all group members. Share those words with the class.

Can you see how using slang in your reports is not helpful?

Chronological Writing

If you write a report chronologically, you record events in the order in which they happened to you. The important part of this message is " . . . as they happened *to you*." The only point of view you should record in your report is your own.

Consider this report excerpt:

> After dinner, I told the residents to tidy their rooms in preparation for an inspection before lights out. I was in the kitchen for about 30 minutes and was just finishing up the cutlery count when I heard a yell from upstairs. I locked the kitchen door and ran upstairs. Mike and Altaf were in the hallway, and they were punching each other. Li was standing in his bedroom doorway. Twenty minutes ago Mike had called Altaf a "pansy." The two youth traded insults from their bedrooms for about 10 minutes, and then Altaf decided to challenge Mike to a fight. I broke up the fight and sent each youth to his room.
>
> I then called the director for further instructions.

At first reading this report seems to make sense; however, how could the report writer have heard Mike call Altaf a pansy if the writer was down in the kitchen doing a cutlery count? How would the writer know how long the boys had been insulting each other?

Here's how the writer should have written the report:

> After dinner, I told the residents to tidy their rooms in preparation for an inspection before lights out. I had been in the kitchen for about 30 minutes and was just finishing up the cutlery count when I heard a yell from upstairs. I locked the kitchen door and ran upstairs. Mike and Altaf were in the hallway, and they were punching each other. Li was standing in his bedroom doorway. I broke up the fight and sent each youth to his room.
>
> I then interviewed Li to see if he had witnessed what led up to the fight. He told me that 20 minutes before the fight, Mike had called Altaf a "pansy." According to Li, the two youth traded insults from their bedrooms for about 10 minutes, and then Altaf decided to challenge Mike to a fight.
>
> I then called the director for further instructions.

A useful technique to help you organize your report into chronological order is to imagine that you had a video camera attached to your shoulder recording what you experienced. Write your reports according to what the camera would have recorded.

Report Organization

Unlike most essays, which use only paragraphs as organization, reports are organized with headings and subheadings. Reports also often have lists, graphs, illustrations, and other visual elements.

Reports are organized this way so that the reader can quickly scan the report to find the parts that are of the most interest. Most people find reports easier to read than essays, because the information is broken into manageable chunks with headings that tell the reader exactly what to expect in the text that follows.

Here are two versions of a report written by a front-line worker in a homeless shelter for young males. The shelter has just opened, and the director of the shelter has asked the front-line worker to find the location of neighbourhood banks, drugstores, and community centres. The information will be used to compile a resource pamphlet for staff members and clients.

Version 1

> To: Chloe Legault
> From: Robin Field
> Date: January 2, 2009
> Re: Neighbourhood Resources
>
> I've been able to locate many community resources for our clients. I think you'll be pleased with the number of banks, drugstores, and community centres that operate in this

neighbourhood. There are four banks within walking distance of the shelter: the Bank of Montreal, Bank of Nova Scotia, the Canadian Imperial Bank of Commerce (CIBC), and the United Credit Union. The Bank of Montreal and the Bank of Nova Scotia are open Monday to Friday from 10:00 AM to 4:00 PM. The CIBC is open Monday to Friday 10:00 AM to 5:00 PM and is also open on Saturday from 12:00 PM to 2:00 PM. The United Credit Union is open Monday to Friday from 7:00 AM to 6:00 PM and is also open on Saturday from 10:00 AM to 3:00 PM. The Bank of Nova Scotia and the CIBC have ATMs accessible 24/7. There are two neighbourhood drugstores. One is a small independent drugstore called "Randall's Drugs," and the other is a Shopper's Drug Mart. Both are open seven days a week. Randall's carries some dry goods and milk and other drinks. Shopper's has some dry goods; a small selection of fruit, vegetables, cheese, and bread, as well as milk and other drinks. It also has a Canada Post outlet. There are two community centres close to our shelter. The Cartierville Community Centre has basketball courts, an indoor gym, classrooms, a kitchen, and a meeting room. It offers a variety of classes that might interest our clients such as wood-working, tae kwon do, and chess. The second nearby community centre is the Ridgewood Centre. It has an outdoor pool, basketball courts, a baseball field, a small weight room, and classrooms. The centre offers many interesting classes such as fencing, calligraphy, and film appreciation. I think this is a great neighbourhood for our shelter. There are many useful, close-by resources. I would be happy to design a pamphlet describing these resources so we can hand them out to any of the youth who drop in.

Version 2

To: Chloe Legault
From: Robin Field
Date: January 2, 2009
Re: Neighbourhood Resources

Introduction

As per your request, I've been able to locate many community resources for our clients. I think you'll be pleased with the number of banks, drugstores, and community centres that operate in this neighbourhood.

Neighbourhood Banks

There are four banks within walking distance of the shelter:

Bank of Montreal
Hours of operation – Monday to Friday, 10:00 AM – 4:00 PM

Bank of Nova Scotia
Hours of operation – Monday to Friday, 10:00 AM – 4:00 PM
ATM on premises

Canadian Imperial Bank of Commerce
Hours of operation – Monday to Friday, 10:00 AM – 5:00 PM
Saturday, 12:00 PM – 2:00 PM
ATM on premises

United Credit Union
Hours of operation – Monday to Friday, 7:00 AM – 6:00 PM
Saturday, 10:00 AM – 3:00 PM

Neighbourhood Drugstores

There are two neighbourhood drugstores. One is a small independent drugstore called "Randall's Drugs," and the other is a Shopper's Drug Mart.

Randall's Drugs
Hours of operation – Monday to Thursday, 8:00 AM – 6:00 PM
Friday, 8:00 AM – 9:00 PM
Saturday, 10:00 AM – 4:00 PM
Sunday, 12:00 PM – 4:00 PM

This store has a small dry goods area and sells milk and other drinks.

Shopper's Drug Mart
Hours of operation – Monday to Friday, 8:00 PM – 9:00 PM
Saturday, 8:00 AM – 5:00 PM
Sunday, 10:00 AM – 4:00 PM

This store has some dry goods; a small selection of fruit, vegetables, cheese, and bread, as well as milk and other drinks. It also has a Canada Post outlet.

Neighbourhood Community Centres

There are two community centres close to our shelter.

Cartierville Community Centre
The centre has the following facilities: basketball courts, an indoor gym, classrooms, a kitchen, and a meeting room.

The following classes might interest our clients: woodworking, tae kwon do, and chess.

Ridgewood Centre
The centre has the following facilities: an outdoor pool, basketball courts, a baseball field, classrooms, and a small weight room.

The following classes might interest our clients: fencing, calligraphy, and film appreciation.

Conclusion

This is a great neighbourhood for our shelter. There are many useful, close-by resources. I would be happy to design a pamphlet about these resources to give to any of the youth who drop in.

Both reports contain the same information, but the second version has been organized by using headings and subheadings. Depending on where you work and what kind of report you are writing, you may have access to standard report templates. If you do have access to templates, when you write a report you would open the appropriate computer template (e.g., incident report template) and plug your information into the designated sections.

In other cases, you will have to design and organize your own report. Smaller facilities (e.g., shelters, drop-in centres, halfway houses) do not often have the funds to buy customized computer software. Here are some guidelines for designing and organizing a report.

Headings

Depending on the length of your report, it may have headings, subheadings, and even sub-subheadings. The report may also include lists.

Visual Design

In order for the reader to understand your report's organizational structure and how the ideas in your report relate to one another, your report's heading levels should look different. In the sample report, the main headings (i.e., "Introduction," "Neighbourhood Banks," "Neighbourhood Drugstores," "Neighbourhood Community Centres," and "Conclusion") are all formatted in 12-point, boldface type. The subheadings (i.e., the names of the banks, drugstores, and community centres, are formatted in 12-point italic type.

The general rule for designing headings and subheadings is to give the bigger divisional markers more visual impact than the lower headings.

For example,

> **Report**
>
> **Heading**
>
> **Subheading**
>
> *Sub-subheading*

Parallel Structure

Your headings should be parallel in structure.

Not parallel	Events before the Incident
	Events during the Incident
	What Happened After

Parallel	Events before the Incident
	Events during the Incident
	Events after the Incident

> **Note:** Headings do not require punctuation of any kind. Specifically, do not put a colon at the end of a heading.

White Space

Use white space effectively in your reports. White space is any area on a page that does not contain any writing. The following tips will help you use white space wisely.

- Leave more space between two sections than you leave between a heading and the text that follows it.

Incorrect

> . . . and that's when Matthew Fry grabbed the knife from Nelson's hand.
>
> **Events after the Incident**
>
> Matthew Fry was able to restrain Nelson, and I called 911. Police officers arrived within five minutes, and they arrested Nelson and took him to the Brown Street Police Station.

Correct

> . . . and that's when Matthew Fry grabbed the knife from Nelson's hand.
>
> **Events after the Incident**
>
> Matthew Fry was able to restrain Nelson, and I called 911. Police officers arrived within five minutes, and they arrested Nelson and took him to the Brown Street Police Station.

- Do not start a new heading at the bottom of a page if you cannot write at least one line under it.

Incorrect

> . . . Michael saw Nelson pull a knife on Tom and heard him yell, "I'm going to make you pay for stealing my girlfriend, you jerk!" Nelson was then distracted by Michael's scream, and that's when Matthew Fry grabbed the knife from Nelson's hand.
>
> **Events After the Incident**

Correct

> . . . Michael saw Nelson pull a knife on Tom and heard Nelson yell, "I'm going to make you pay for stealing my girlfriend, you jerk!" Nelson was then distracted by Michael's scream, and that's when Matthew Fry grabbed the knife from Nelson's hand.
>
> **Events After the Incident**
>
> Matthew Fry was able to restrain Nelson, and I called 911. Police officers arrived within five minutes, and they arrested Nelson and took him to the Brown Street Police Station.

Typefaces and Styles

The report writing you do must look professional. This means that you should use standard typefaces that are widely known.

Some typefaces, such as the following, are unsuitable for the task:

Incident Report

Incident Report

Some typefaces, such as the following, are hard to read:

Incident Report

Incident Report

Incident Report

Stick to a simple serif typeface such as

Times New Roman (Incident Report)

or

Garamond (Incident Report)

Or a simple sans-serif type face such as

Trebuchet MS (Incident Report)

or

Arial (Incident Report)

Note: Many readers do not like to read text that is all capitalized. Words made of capital letters are hard to read and are often equated with shouting at the reader.

Additional Considerations When Writing Reports

Reports are sensitive documents. Here are three guidelines to help you ensure that your reports are secure and professional.

Respect Confidentiality

It is critical that the reports you write are not read by unauthorized people. Follow these procedures to ensure that your reports remain secure:

1. Lock reports into cabinets.
2. Lock the door to your workspace, if possible, when you are not in it.

3. Close logs or turn over reports when someone else is in your workspace.
4. Keep the work you bring home, if possible, in a locked or secure area.
5. Keep the work you transport to another location in a safe environment (e.g., in your car trunk) or under your supervision (e.g., keep your briefcase with you).
6. Do not discuss a client with anyone other than those involved in the client's case.

Use a Pen for Handwritten Reports

Do not write your reports in pencil to ensure that your work cannot be altered by someone else. Use a black- or blue-ink pen when handwriting reports. Coloured inks often fade and do not reproduce well on photocopiers.

Sign All Reports or Case Notes

Sign your work to authenticate it and to let other readers know who to approach for clarification.

Correct Errors Properly

Put a single line through any corrections or changes you make, and initial the changes. This guideline applies to handwritten changes to a typewritten document as well as handwritten changes to a handwritten document. If the information on a report is crossed out in such a way that the original information is illegible, the reader may infer that the writer attempted to alter events to cover up mistakes.

Chapter Summary

- Your reports must be objective and factual.

- Workers in the community services field write reports in the active voice.

- You should avoid jargon, slang, and clichés in your report writing.

- You can organize your reports using headings, subheadings, and white space.

- You must ensure you respect confidentiality and accountability.

Chapter Exercises

EXERCISE 1.1 Editing for Objectivity, Active Voice, Slang, and Clichés

Rewrite the following report excerpts. Make sure they are objective and written in the active voice. Remove any subjective wording, slang, or clichés. You may add or delete information as necessary.

1. It was easy as pie to get the delinquents to behave, because most of them follow their leader Bill like sheep. I simply told Bill that if his friends continued to argue about the TV, I would deduct one point from each of their daily totals, and I would ban them from the TV room for two days. The little jerks sure behaved after that.

2. I couldn't get the street bum to listen to me, so I called the police on the low-life scum. The cop who responded to my call was pretty good-looking. He talked to the drunk and then left me his card in case there was any future trouble.

3. Marcin was given a consequence for his inappropriate language at the assessment hearing. He was told to apologize to the parole officer. He was also advised to control his temper.

4. This afternoon, at about 1500h, Abdi asked me for an appointment. He was agitated and paced the floor the entire time he was in my office. Abdi told me he was being physically threatened by Kevin. Since the dude is one of my favourite clients, I believe him. I will be speaking to Kevin on my next shift.

5. My afternoon in court really sucked. The judge must have gotten up on the wrong side of the bed. She heard the defence lawyer's opening statement and then recessed court for the day. So much for my testimony. I guess I'll have to go back to court tomorrow.

6. This morning I assessed a cute new client, April, for admission into our anger management program. April's parents are both long-term drug users, and April has been using drugs since she was 11. Poor thing. She was recently convicted of robbing a corner store and is currently completing community service orders.

7. While on shift this afternoon, I witnessed a verbal altercation between two residents. Sheila and Tia got into a cat fight about who had better hair. Those two get along like oil and water. I was able to de-escalate the argument, but I am concerned that the two women are still angry with each other. I advise the evening staff members to watch them like hawks.

8. The youth was arrested and taken down to the Bethany Street Police Station. Charges were placed against the youth, and he will see the judge in the morning.

EXERCISE 1.2   Writing Short Reports

Use the following scenarios to write short reports. Refer to the report writing elements discussed in this chapter to help you with your reports' organization, contents, and layouts.

1. Write a report about your college experience so far. Consider your course load, your homework, the people you have met, and your teachers as possibilities for sections of the report. Add any sections you think would be relevant.

2. Write a report about your favourite restaurant. Consider the food, the atmosphere, the staff, and the prices as possibilities for sections of the report. Add any sections you think would be relevant.

3. Write a short report about an influential person in your life. Consider the person's physical appearance, character traits, and relationship to you, as well as how that person has changed you as possibilities for sections of your report. Add any sections you think would be relevant.

4. Write a short report about a vacation you have taken or would like to take. Consider location, cost, and attractions as possibilities for sections of the report. Add any sections you think would be relevant.

5. Write a short report about a disappointing experience you have had. Consider writing about a terrible meal, poor service, a bad job, or a boring date. Before writing, decide what points you would like to make about your topic. Develop each point into one section of your report.

EXERCISE 1.3 Using Parallelism

Rewrite the following sentences so they are parallel in structure.

1. While at the cottage, I like to swim, canoe, and reading.

2. Resident Franklin was not cooperative during the interview; he swore repeatedly, drumming the desk with his feet, and there was no eye contact made.

3. The parole officer asked his client to sit down, explaining the parole hearing process, and he also told his client to not be nervous.

4. As an outreach officer for street youth, my job is giving the youth food, asking if they need blankets, and I try to find out if they need medical attention.

5. Eating well, if you get a good sleep, and try to relax are three ways to be ready for an exam.

6. To de-stress after a hard day I like to ride my bike for an hour and I've also been known to eat ice cream.

7. When you are writing a funding proposal, you should clearly state your program goals, there should always be a budget included, and specifically outline your expected outcomes.

8. At the Social Service Worker Conference, we were able to network, practise new interview techniques, and we had some great speakers.

9. Our anger management program should include role play, we should have participants do self-evaluation, and small group discussions.

10. The local MPP is visiting our group home next week to see our new addition, talk to us about getting more funding, and she will also be presenting us with an achievement award.

In-House Reports

LEARNING OBJECTIVES

After reading this chapter and completing the exercises in it, you will be able to

1. Understand the narrative process used to write in-house reports.

2. Integrate the report writing basics learned in Chapter 1 into in-house reports.

3. Improve your narrative writing style by completing a variety of in-house reports.

Introduction

This chapter discusses the *in-house reports* you will write in the community services field. In-house reports are reports that document the on-going progress of your clients.

You will likely have many meetings with your clients to discuss such things as their progress in a new residence, the programs in which they may be enrolled, the feelings they may be having, their day-to-day experiences, and the incidents in which they may have been involved. The content of all your encounters with your clients, regardless of how casual your contact may seem, should be documented as part of your clients' files. These files are invaluable to all professionals involved in helping your clients, and they will be read by many people.

Some reports you will write may include

• Intake interview notes	• Case notes	• Incident reports
• Behavioural contracts	• Progress/Goal-setting notes	• Plans of care
• Daily logs	• Assessment reports	• Serious occurrence reports

SENTENCE POLISHING SKILL: DIRECT SPEECH IN REPORTS

When you report *direct speech*, you record the exact words of a speaker.

Example of Direct Speech

Francine reached for the last donut and said, "I really shouldn't eat this, but I am going to anyway."

Example of Indirect Speech

Francine reached for the last donut and said that she really shouldn't eat it, but she was going to anyway.

You will want to use direct quotes for a variety of reasons in your writing: you may be quoting a source for a school paper; you may be writing an incident report that describes a fight during which threats were uttered; or you may be including persuasive testimonials in a funding proposal.

The following text explains the guidelines for punctuating direct speech.

Rules for Recording Direct Speech

1. If you are recording a complete sentence that someone said, put a comma before the statement, use open quotation marks at the beginning of the quote, end the quote with the appropriate punctuation mark (period, question mark, or exclamation mark), and use closed quotation marks at the end of the quote.

   #### Example
   Mustafa then yelled at the child and youth worker, "You'll pay for cancelling my leave time."
   Note: The closing quotation mark is placed after the period.

2. If you are recording only a few words of someone's speech, you do not need a comma before the quote.

   #### Example
   I have recorded Mustafa's threat of "you'll pay" and scheduled him for a meeting with our house director this afternoon.

At the end of this chapter, you will find an exercise to help you practise recording direct speech.

In-House Reports

As noted in the introduction to this chapter, you might be expected to write a variety of in-house reports. The following text provides descriptions of five types of in-house reports.

Intake Interview Reports

Once a client is assigned to your facility, shelter, program, or care, it is standard procedure to have an intake interview with the client to introduce yourself and your services to the client, to explain procedures, to gather the client's history, and to make plans for the client.

Most facilities have forms for this procedure, because the process includes gathering basic information about the client, such as personal statistics; medical, family, and criminal history; social history; and previous or on-going treatment plans.

As well, these intake interview forms have a section in which you will write a narrative that records the conversation you have with your client. You will be asking the client a variety of questions to establish rapport and understand his or her specific needs and expectations.

Here is an excerpt from an intake interview report written by a worker at a shelter for women fleeing from domestic abuse. The report records a section of the intake interview between the worker and a female adult staying at the shelter for the first time.

Intake Interview

Monday, July 28, 2009

Today, I met with Harriette Feeny to discuss her residency at A Safe Place. This is Harriette's first time at the shelter. She brought with her a two-year-old son (Austin) and a four-year-old daughter (Leah). I began the interview by introducing myself and telling Harriette about the shelter. I gave her a few minutes to read our policies and procedures pamphlets and asked her if she had any questions. She had one. She asked me if her 17-year-old son Eric would be able to visit. I told her he could not visit at the shelter, but we could arrange for a safe meeting place for them and escort her on the visits, if she wanted. She seemed greatly relieved at this. She said that not being able to see her oldest son had been her only concern when she was deciding whether to use the shelter. During our meeting, Harriette said little. She seemed tired, and when I asked her to tell me about her plans for returning to work and for her children, she asked if we could meet another time to discuss this. She then asked me if she could see her room and have a nap. I felt that it would be best if we continued the interview the following day. I took Harriette to the resource worker so Harriette could be given a room. I also asked the daycare staff to keep Harriette's children for an additional hour or so while Harriette napped. I have scheduled a meeting with Harriette for tomorrow (July 29) at 9:30 AM.

ACTIVITY 2.1 ## Writing an Intake Interview

Imagine you are a worker in an open-custody facility for youth in conflict with the law. You must interview a new client, a 15-year-old male on probation. In your interview, you want to welcome the youth and let him know he is entering a safe and positive environment. You also want to cover these topics: house rules, the youth's anger management program, and the youth's high school enrolment at the local high school. Write your intake interview report, including all the necessary details.

Daily Logs

Daily logs are the recordings of what went on every day, for every shift, at a facility, treatment centre, drop-in centre, shelter, or home. All front-line workers are expected to write a summary of their shifts before they go off duty.

Daily logs serve many purposes. In an ideal situation, on-coming shift workers have time to sit down with workers ending their shifts to discuss what happened on the previous shift; however, this is not always possible. A properly completed daily log is a necessary informational tool for the on-coming staff. Logs provide historical reference; they alert staff members of any problems or concerns; and they allow staff members who might never work together to communicate with one another.

A daily log book, in the form of a binder or notebook, is always kept in the office. Staff members are usually required to start their shifts by reading the log from the previous shift (or shifts, if they have had some days off). They then make regular entries to the log book throughout their shifts when possible and complete their daily log submission at the end of their shifts.

Sometimes it is not possible for staff members to make regular entries in the daily log throughout their shifts. On the occasion that staff members are not able to make any entries into the daily log until the end of their shifts, they must take care to record events as they happened, so the chronology of the shift is correct. Staff members should write a quick rough draft before they write in the log book. This extra few minutes (as tired as workers may be at the end of their shifts) ensures a correct log book entry.

Here is a sample of a daily log entry from a worker at a halfway house for young male offenders:

Date and Shift: Friday, October 27, 2008, day shift (8:00 AM – 4:00 PM)
Worker: Alexandra Newman
Report: At 8:00 AM, I clocked in and read the night log. In his log, staff member Alan Cheung reported that two residents, Collin and Victor, had been arguing over kitchen duties after supper. Alan advised that the day workers should closely watch the two youth, as their argument could easily become physical.

Minor Incident/Intervention

At 11:35 AM, Collin and Victor were in the kitchen helping me with lunch. They began to argue about whose turn it was to make the sandwiches. The argument seemed about to escalate, so I had both youth take a seat opposite each other at the table. I served them coffee and asked them to take turns explaining their frustrations. After listening to them, I realized that both Collin and Victor thought they were doing more work than the other. I proposed that the three of us draft a work schedule. They agreed with the solution, and a two-week schedule is now posted on the kitchen door. Both residents have promised to see me if they have frustrations with the schedule or other work issues.

Recommendation

I think we should continue to update the schedule, with input from both residents. I am willing to take responsibility for this.

Conclusion

The rest of my shift was uneventful. A furnace repair person came in at 2:40 PM to clean the furnace. I accepted delivery of our weekly vegetable order at 3:05 PM.

Off Shift: Matthew arrived for the 4:00 PM shift, and I clocked out.

ACTIVITY 2.2   Writing a Daily Log

Imagine you are a front-line worker at a 24-hour support centre for homeless adults. Your shift is from 2000h to 0800h. Write a daily log that includes the following:

- A client arrives. He is bleeding from a deep cut on his head, and you call an ambulance to take him to the hospital.
- You play cards for an hour with two regular clients who come in to get out of the rain.
- Police officers visit you to ask if you have seen a regular client that evening. He is wanted for questioning because he witnessed an assault.
- A newspaper reporter comes into the centre and asks to interview you for her paper. You decline, give her your supervisor's work number, and advise her to call in the morning.

You must invent all the necessary details (including times) to complete this daily log.

Case Notes

From time to time (e.g., weekly, bi-weekly, or even daily), you will meet with clients to respond to concerns or to suggest programs or other steps for improvement. Case notes are your recordings of what was discussed, your impressions of the client's emotional or physical state, and any plans you and the client develop. These case notes provide a valuable reference for everyone involved with your client.

Case notes may be written in a binder or notebook. In some facilities, they are written on pre-printed forms that are filed in the client's folder.

Here is a case notes sample from a meeting with a young offender living in an open-custody halfway house. A child and youth worker monitoring his client's high school attendance and performance wrote the report.

Case Notes: Robert W.

Date: October 23, 2009

Today, at 4:00 PM, I met with Robert. He has been living at McCaully House since September 7, 2009, and has been attending Field Glen High School. He is in grade 10 and is taking applied-level courses. Robert started our meeting by telling me he is enjoying his math class very much but has fallen behind in both his chemistry and English classes. I had already received a call from his English teacher, so we discussed that first.

According to the teacher, Robert is inattentive in class and is, at times, disruptive. When I asked Robert about this, he admitted to this behaviour and told me he was "really bored" in the class. I asked Robert what he planned to do about the problem. He said he plans to start sitting at the front of the class and will ask permission from the director of McCaully House to attend the after-school homework club on Tuesdays and Thursdays.

Robert did not want to talk about the English or chemistry class any more, so we began to talk about his math class. He is getting B grades and has tutored a fellow student on a few occasions. One of Robert's goals for his self-improvement plan is to contribute to his environment, both at home and in school. I suggested that Robert should write a journal entry about how this tutoring has helped both him and the student he tutors.

I asked Robert to make an action plan for his chemistry class and arranged to meet with him next week to discuss both that and his English class.

Our visit ended at 5:00 PM.

ACTIVITY 2.3   Writing Case Notes

You are a front-line worker working with a program that offers services to homeless youth. Write case notes about your interview with a male youth who has used the program off and on for three years. Your interview was about the youth's plans to attend college. The youth has just signed on to social assistance and is looking for an apartment to rent.

Progress Notes

As well as general case notes, you may need to write progress notes. If your clients are enrolled in programs (e.g., anger management, personal hygiene, education), medical treatment (e.g., drug or alcohol, mental health), or counselling (e.g., grief, anxiety, depression), you will want to conduct regular meetings to discuss their progress. For samples of progress notes, please see the progress notes below and Template 2.1 on page 31.

You could include your progress notes in your case notes, but the benefit of keeping separate notes for progress interviews is that they are easily accessible for reference if there is a problem.

Here is a sample of progress notes written by a front-line worker in a group home for adolescent women:

Progress Notes: Serenity House
Client: Keetha S.
Date: October 23, 2009
Worker: Tara Peters

Introduction

Today, I met with Keetha for our weekly progress report. We discussed her impressions of the anger management class she attended on October 20, 2008.

Background

Keetha's psychologist referred her to an anger management program. The program is six weeks long and is targeted specifically to teens who have been physically abused by their parents. Keetha agreed to try one class, with the view to completing the six-week program if she felt it could be helpful.

Interview

I began the interview by asking Keetha if she felt the session helped her. Keetha told me that although she enjoyed hearing the other participants speak, she didn't think she would ever be comfortable talking to the group. She went on to say that one girl, who Keetha described as "stuck-up and negative," was very intimidating. I asked Keetha if she learned anything at the session. She said that the facilitator told the group that anger was natural and even healthy at times. Keetha told me that this was the first time anyone had ever said that to her. She said she agreed with that statement and that she liked the facilitator. Keetha went on to say that the facilitator gave the group a calming exercise to try. Keetha told me she plans to try the exercise the next time Barbara gets on her nerves. (Barbara is a resident with whom Keetha seems to get in many verbal arguments.) I will be watching to see if she is able to follow this plan. I hope to be able to praise Keetha if she can avoid a confrontation with Barbara.

Conclusion

Since Keetha found the first session beneficial, she has agreed to continue to attend the weekly sessions. I will meet with Keetha every week after her session to help her process what she has learned.

ACTIVITY 2.4   Writing Progress Notes

You are a front-line worker in a group home for youth who have substance abuse issues. Interview a resident about her progress in an anger management program. She hates the program and would prefer to start one-on-one counselling. Write progress notes about this client. Invent all the necessary details to complete the notes.

Incident Reports

As a community services worker, you will be working with diverse groups of people, many of whom will have social, behavioural, and emotional challenges. Your clients may be coping with agitation, dealing with stress, or working through crises. You can expect incidents such as threats, violence, self-injurious behaviour, or accidental injuries. These incidents need to be fully documented in incident reports. Incident reports are the in-house reports most likely to be shared with outside agencies, especially in the case of criminal behaviour. For samples of incident reports, please see the report below and Template 2.2 on page 34.

> **Note:** Many halfway houses, centres, drop-in shelters, and other criminal justice system facilities have computerized templates for incident reports, which makes it easy for the writer to organize the report's details. The sample report below uses a simplified incident report format suitable for the purposes of this textbook.

Here is a sample incident report:

Date, Time, and Place of Incident
This report details an incident that occurred on January 5, 2009, at 7:15 PM at McNeil House.

Nature of Incident
Verbal altercation between a front-line worker and a client.

Name(s) of Person(s) Involved
Elif D. – client involved in altercation
Katherine Bowie – staff member
Ian W. – client who witnessed the altercation
Harry Newault – staff member

Events Leading to the Incident
Elif D., a regular client of our drop-in centre, arrived at the centre on January 5 at 7:15 PM. She signed in at the front desk and then went into the TV lounge. At that time, staff member Katherine Bowie and client Ian W. were watching a nature show. Elif took the remote control and changed the channel.

Events during the Incident
Ian told Elif to change the channel back, and Elif refused. I (Katherine Bowie) then asked Elif to change the channel back. She refused again. I approached Elif, put out my hand, and asked for the remote control. Elif refused to comply. At this time, I was approximately a metre from Elif, and I smelled alcohol on her breath. I asked Elif if she had been drinking that evening. She replied, "What's it to you? I can have a friggin' drink whenever I want." As it is against centre policy for clients to attend under the influence of alcohol, I asked Elif to leave. She became more agitated and started to scream obscenities at me. I asked her to leave again. Elif complied, and as she left, she continued to scream obscenities, which Harry Newault also heard from where he was working at the front desk.

Action Taken in Response to the Incident
As per McNeil House policy, Elif is banned from the centre for 30 days, and I will place a copy of this report in her file.

ACTIVITY 2.5 # Writing an Incident Report

Write an incident report for the following scenario. Use the layout of the sample report as a guide.

You are a youth worker at a YMCA. You are in charge of an after-school program for teens. Imagine that a youth got into a verbal argument (escalating toward violence) with another youth over a basketball game. You were able to calm down both youths before they became physical. You told both youths that the incident would be recorded and that there would be no further action as long as the youths continued to behave.

Invent all the necessary details to tell the complete story.

Chapter Summary

■ In-house reports explain, describe, or discuss client progress and daily occurrences in any given facility.

■ A variety of community services professionals read in-house reports, so your reports must follow basic report writing principles.

■ Write your reports using the narrative writing-style.

Chapter Exercises

EXERCISE 2.1 # Writing Intake Reports, Case Notes, and Progress Reports

Write reports for the following scenarios. For each scenario, you will need to invent details to complete your reports.

1. You are a child and youth worker in an elementary school. You have been asked to conduct an intake interview for a 10-year-old girl who is having severe aggression problems. She has also exhibited self-injurious behaviour. In the interview, you get to know the girl and find out her triggers for negative behaviour. You also tell her that you are there to help and support her.

2. You, a housing officer, have scheduled to meet with a client who has been living in supervised independent housing for the last two months. He begins your meeting by expressing concerns about his neighbours in the building. He is a recovering alcoholic, and the neighbours continue to offer him alcohol. He has refused so far, but he does not think he can hold out, even for one more day. You tell him you will immediately deal with the neighbours, as it is against the rules to have alcohol on the premises. Write the case notes for this scenario.

3. You are a front-line worker in a closed-custody halfway house for female youth in conflict with the law. You interview a client who has just completed a conflict management program. She did well in the program and seems ready to move forward. You get her opinion of the program and learn about her success in it. You also, with her help, develop an educational plan for her for the next few months. Write the progress notes for this scenario.

4. You are a front-line worker in a halfway house for adult male offenders. Interview a resident about how he broke curfew and was AWOL from the house for two hours. Write the case notes about your interview.

5. You are a social services worker with a client who is living in a foster home and receiving home-schooling because she has been expelled from three high schools. Interview her about her progress in school. Write the progress notes for this interview.

EXERCISE 2.2 Writing Daily Logs and Incident Reports

1. Write a serious occurence report entry using the following information.

   • You work at a group drug rehabilitation home for adult males.
   • You worked the overnight shift from 2000h to 0800h.
   • At 2020h you noted that the kitchen had not been cleaned up from the evening snack.
   • You found a used syringe in the downstairs bathroom. You did not touch the syringe. You locked the bathroom door and put an "Out of Order" sign on it. You called the director, who told you he would be in at 0800h to deal with the issue. He told you to keep the bathroom door locked.
   • You received a call from Grace Mercy Hospital asking if you had room for a new resident. You told the special care nurse that you would have the director phone the hospital the following morning.

   Invent all the necessary details (including times) to complete this daily log.

2. Write an incident report for the following scenario. You are a worker at a group home for adolescent male offenders. On January 5, two residents, Maurice N. and Tyrone K., were involved in a fist fight in the TV room over which show to watch. You and your colleague, Steffany Trites, broke up the fight, but not before two lamps were broken and Maurice N. received a black eye. Follow-up interviews with the boys revealed prior friction over Amanda S., a girl at school. Tyrone K. has a history of fighting, both at school and in the group home. Your report will be placed in each youth's file.

   Invent all the necessary details to write a complete report.

3. You are a front-line worker in a halfway house for female young offenders. Write a daily log to record the following events.

   • You worked an evening shift.
   • When you conducted the cutlery count, two knives were missing.

- A search resulted in a staff member finding the knives in a resident's room.
- The police were called to arrest the girl for parole violation.
- Two other residents had a fight about TV programs.
- Fighting over the TV has been an ongoing problem for these two residents, so you suspended their TV privileges for one week.
- The rest of your shift was uneventful.

4. You are a correctional guard at a minimum-security correctional facility. Write a daily log to record the following events.

- For the first hour of your shift, you gave college students a tour of your wing at the facility.
- Some of the inmates became agitated as a result of the tour and shouted insults and inappropriate comments at the students.
- One inmate grabbed a female student by the shoulders and tried to kiss her.
- You confined the man to his room and ended the tour.
- You called the inmate's parole officer, and he will be meeting with the inmate the next day.
- The rest of your shift was uneventful.

5. You work at a facility for troubled youth. Write a daily log to record the following events.

- You started your shift by putting on a movie for the youth.
- You helped two of the clients make popcorn for the group.
- Halfway through the movie, one of the clients stormed out of the room.
- You followed the youth and had a conversation with him, during which he told you that two of the people also watching the movie were whispering to him throughout it.
- They accused him of being gay and threatened to beat him up when he went to bed.
- You stopped the movie and had a conversation with the two youths.
- You took away their TV privileges and put them on probationary warning.
- At about 2:00 AM, it began to rain and the roof leaked in the kitchen and front hallway.
- You put buckets under the leaks.
- The rest of your shift was uneventful.

EXERCISE 2.3 **Reporting Direct Speech**

Add the punctuation and quotation marks necessary to correctly record the direct speech in the following sentences. The direct speech is underlined for clarity.

1. The youth was despondent about her visit with her mother. In our interview, she said I wish my mother had given me up for adoption when I was a baby.

2. During the fight, Noel slammed a chair into the wall and yelled I'm fed up with this crap. I'm outta here. He then left the shelter.

3. Johnson confessed that his anger management program was not going very well. He claimed to have feelings of anxiety whenever the group had to do role play.

4. After many interviews with Cathy, a sex trade worker who visits the women's drop-in centre regularly, I still cannot convince her to enroll in the methadone program. She has taken the program before and told me the program just leaves her wanting <u>the real thing</u> even more.

5. I was pleased about my performance review. My boss told me I am an outstanding employee who is <u>a positive role model for both clients and staff members.</u>

6. Nicky and Angel were involved in another verbal altercation last night. Angel had taken clothing out of Nicky's room, and Nicky was heard to say <u>If I catch you in my room again, I'll cut you good.</u>

7. According to the newspaper article I read, the halfway house director was disappointed that the neighbours are not more supportive of the youth in the home. He was quoted as saying <u>It's too bad we get such resistance about our youth. We have been operating in this neighbourhood for seven years and have been no trouble whatsoever.</u>

8. The security officer asked the agitated youth to step outside the shelter. The youth responded by breaking down and sobbing uncontrollably. The youth went on to say that he had just found out that his girlfriend was a <u>lying, cheating, manipulator,</u> and he just wanted to <u>show her who was boss.</u>

TEMPLATE 2.1 Sample Progress Notes

Progress Notes: Curtis Royale

Date of Meeting: November 30, 2008

History

Curtis has been a client of mine for six months. We meet once a month to discuss his progress with the plan we developed when he first became a client of Marmorelle Family and Social Services. Curtis is a recovering cocaine addict with a minor criminal history. Curtis has committed to the following:

1. Attendance at night school for high school upgrading

2. Attendance at a 12-step addictions class

3. Part-time work stocking shelves at a local clothing warehouse

Curtis plans to complete his high school upgrading and then begin forklift-operator training.

Progress Interview

We began our interview at 11:00 AM with a discussion about how Curtis is progressing in his two night-school courses. Curtis reported to me that he still enjoys both classes and is maintaining a C- in both. He will complete the classes at the end of December, and he asked me if I could help him register for two more classes for the winter term. I agreed to help.

I asked Curtis about his attendance at the 12-step addictions classes. He told me that he attends once or twice a week and gets a lot of support from the other attendees. He told me his sponsor is particularly good at understanding Curtis's stressors. As an example of this, Curtis reported to me that recently a friend of his was released from Marmorelle Correctional Facility. Curtis was looking forward to re-connecting with the friend, but knew the friend would want Curtis to party with him. Curtis told his sponsor about the friend's release and, since then, the sponsor has been calling Curtis every night at 10:00 PM to check in with him and to have a friendly chat. Curtis told me this has been a "lifesaver" for him.

Curtis told me he does not particularly enjoy his job, except for the physical aspect of it, but that he has met a few decent people.

Overall, Curtis told me he is happy with his new life and is looking forward to training as a forklift operator and putting his past behind him. I asked Curtis if he still wanted the support of MF&SS, and he replied that he did. He went on to tell me he feels proud of his progress and that our monthly meetings help him realize how far he has come.

Our interview concluded at 11:50 AM, and we agreed to meet again on December 28. We will also continue our weekly progress telephone interview.

Assessment

I am pleased with Curtis's progress. He is committed to improving his life and his request to take more courses indicates he is independently making positive choices for change.

TEMPLATE 2.2 Sample Incident Report

Phoenix Place

Incident Report

Incident Type: Absent without Leave

Individuals Involved: Client – Margo Struthers, Staff members – Noah Clermont and Emily Piper, as well as Police Constable Ned Grainger and Probation Officer Lucy Huang

Date and Time of Incident: Wednesday, July 21, 2008, 11:00 PM

Precipitating Events

At 3:00 PM on the day of the incident, client Margo S. asked me, staff member Noah Clermont, if she could go to an evening movie with a friend. Margo told me that the movie ended 10 minutes after curfew but that she would be home as soon as possible after the movie let out. This has been occasionally allowed for other clients, and Margo was aware of this. However, Margo has been on program probation for behavioural issues (see the Incident Report dated July 15, 2008, in Margo's file), so I denied her request. I did point out that Margo could attend an earlier showing of the movie and be back in time for the 10:00 PM curfew. Margo replied that she would do that instead. She left the residence at 3:40 PM.

Description of Incident

At 10:00 PM curfew, Margo had not returned to the residence. As per house rules, I waited for the 20-minute grace period to run out and then contacted the police and Margo's probation officer. The police dispatcher told me she would send an officer to the residence. I left a voice message with Lucy Huang, Margo's probation officer. Officer Ned Grainger responded to the residence. He met with me, borrowed a picture of Margo for his search, and told me he would have patrol officers on the lookout for Margo. He left the residence at 11:00 PM.

At 11:45 PM, staff member Emily Piper heard a knock on the front door of the residence. Margo S. was at the door. Emily allowed Margo S. in, sat her down in the lounge, and came to the office to get me. I asked Margo why she broke curfew. She explained that when she met up with her friend, the friend begged Margo to go to the late show because the friend's boyfriend would also be there, and she wanted Margo to meet him. Margo agreed and went to the late show. She told me she left the show early enough to catch a bus that would get her home in time to make curfew, but that the bus did not show up. She was forced to take a later bus. I explained that because she broke curfew, police and her probation officer had been contacted. Margo began crying at this but had nothing further to say.

Results of Incident

I told Margo that her probation officer would meet with her the next day to discuss consequences. I dismissed her to her room.

I then informed police that Margo had returned. I called Director Evans who asked me to complete this report before going off shift.

Police/Medical Intervention

As per the above details, police were contacted. No medical intervention was required.

Noah Clermont   *July 21, 2008, 1:45 AM*
_____   _____
      Staff Signature   Date and Time

Multi-Audience Reports

After reading this chapter and completing the exercises in it, you will be able to

1. Apply the organizational skills introduced in Chapters 1 and 2 to write reports meant for multiple audiences.

2. Use persuasion effectively in your reports.

3. Write comprehensive, clear reports that allow the reader to make informed decisions.

Introduction

Some of the in-house reports discussed in Chapter 2 will also be read by a variety of audiences outside of your facility, but their primary purpose is to provide a record of events for in-house use. You will also write a variety of reports for which outside readers are your intended primary audiences. Front-line workers write these reports in response to specific events, proposed programming, or stages in the justice process.

SENTENCE POLISHING SKILL: PRONOUN USE

A pronoun takes the place of a noun. Here is a list of some pronouns:

- I, me
- You
- He/she/him/her
- We/they/us
- It

Example

<u>Bill</u>'s medication has been changed.
Bill is a noun.

<u>His</u> medication has been changed.
His is a pronoun. It takes the place of *Bill*.

Writers use pronouns to add variety to their writing. The effective use of pronouns polishes writing. The first objective of report writing, however, is to tell a clear story. Often the stories you will report as a community services worker involve several people, and you must clearly record the actions of those people. This means that you must be careful when substituting a pronoun for a noun.

Read the following passage:
> Bill's medication has been changed. <u>He</u> has been having trouble adjusting to the new dosage. As a result, <u>he</u> is often agitated. <u>He</u> recently struck another client, Tony, because <u>he</u> accused <u>him</u> of using <u>his</u> MP3 player without asking. <u>He</u> then told <u>him he</u> hated <u>him</u> and didn't want to be friends any more.

The use of too many pronouns in the above passage makes the story unclear. Who used the MP3 player without permission? Who hates whom?

Here is the same passage with fewer pronouns:
> Bill's medication has been changed. He has been having trouble adjusting to the new dosage. As a result, Bill is often agitated. He recently struck another client, Tony, because Tony accused Bill of using Tony's MP3 player without asking. Bill then told Tony Bill hated Tony and Bill didn't want to be friends any more.

The last part of the above passage may seem stilted; however, your goal as a report writer is to get the story straight. If this means that your writing is not elegant prose, so be it.

Writers must be aware of another troublesome area of pronoun usage.

Consider this sentence, which appears in a memo to staff members:

Anyone wishing to pick up an extra shift should see his supervisor.

The pronouns in this sentence are "anyone" and "his." The problem is that using *his* may be inaccurate, unless all the staff members reading the memo are male.

There are a few ways you can avoid making this error. You can pluralize your nouns and pronouns or avoid using gender-specific pronouns.

Examples

People wishing to pick up an extra shift can see their supervisors.

Extra shifts are available; see your supervisor if you would like to sign up for one.

Note: All pronouns ending in *-one* (e.g., anyone), *-body* (e.g., everybody), and *-thing* (e.g., everything) are singular pronouns.

Because "everybody" is a singular pronoun, the sentence below is not correct:

<u>Everybody</u> wishing to pick up an extra shift can see <u>their</u> supervisor.

The use of "their" as a singular pronoun is becoming common, and you will find it is acceptable in many workplaces. It is recommended that you do not use "their" this way, because as a community service worker, you will be writing legal documents with high professional standards.

At the end of this chapter, you will find an exercise to help you practise clear pronoun use.

Your Audience

Your clients will have many needs that require you to liaise on their behalf with professionals from many fields. Here are some possible professionals that you may write to:

- Hospital staff
- Housing officers
- Community and justice services workers
- Social workers
- School staff
- Psychologists
- Police personnel
- Probation officers
- Defence lawyers
- Parents, guardians, caregivers

- Children's Aid Society (CAS) workers
- Shelter staff
- Social service workers
- Child and youth workers
- Psychiatrists
- Counsellors
- Parole officers
- Judges
- Crown lawyers
- Ministry personnel

Reports for Outside Audiences

Here are some situations for which you might be required to write multi-audience reports:

Client Progress

You may write client progress reports that will be used to determine client readiness for programs, independent living, school, training, or employment.

Serious Occurrences

You may write serious occurrence reports (e.g., about an attempted suicide or assault) to police, CAS, parole officers, probation officers, or other case workers. For a sample of a serious occurrence report, please see Template 3.1 on page 49.

Requests for Services

Your clients often do not have the communication skills they need to successfully advocate for themselves. You might write a letter or report to help a client secure housing, get a job, open a bank account, or apply for social assistance.

Admissions to Programs

There are often program opportunities for your clients. Spots are usually limited, and your clients must meet certain criteria to qualify for a spot. Often clients need a case worker referral to participate in programs.

Program Quality Review

You may write reports to ministry departments, advisory committees, or boards of governors to satisfy funding and licensing requirements for your facility. For a sample of a program quality review report, please see Template 3.2 on page 53.

Multi-audience reports follow the same report-writing conventions as those discussed in Chapters 1 and 2; however, you should also consider the following three elements for your multi-audience reports:

1. Format

Your task and purpose will dictate the format of your multi-audience reports. Your reports can be informal, semi-formal, or formal. Table 3.1 shows how these types of reports differ.

For either an informal or a semi-formal report, you will most likely draft the report's format. Your agency may have templates or samples for your reference. When you write formal reports, your agency will usually give you an organization-specific format that you must carefully follow.

2. Use of Detail

For reports you write to audiences outside your agency, you must provide much more background information and many more client history details than you would if the report was being read in-house. For example, imagine that a provincial social services ministry has funded a new construction job skills program. Clients admitted to the program will stay at a facility for two weeks. They will learn job search skills, attend information sessions by local trades employers,

Table 3.1 Informal, Semi-formal, and Formal Reports

Informal Reports	Semi-formal Reports	Formal Reports
1–3 pages	3–10 pages	10+ pages
Have no cover page	Have a cover page; may have a table of contents	Have a cover page, a letter of transmittal, a table of contents, and an executive summary
Are usually organized into sections with headings; can also be written in letter format	Are organized into sections with headings and possibly subheadings	Are organized into sections with headings, subheadings, and possibly sub-subheadings
Start with an introductory paragraph	Start with introductory paragraphs	Start with introductory paragraphs
Use informal business tone; are usually written in the first person (i.e., use "I")	Use semi-formal business tone; are usually written in the first person (i.e., use "I") but could also be written in the third person (i.e., use "the writer")	Use formal business tone; are usually written in the third person (i.e., use "the writer")
End with a concluding paragraph	End with concluding paragraphs	End with concluding paragraphs
Might have an attachment	Often have appendices - letter(s) of support - charts - statistics	Have appendices - letter(s) of support - charts - statistics - testimonials - suggested readings - a reference page

participate in a hands-on "day on the job" session, and be interviewed by potential employers. Your halfway house for male youth has just received an invitation to propose a suitable client for the program. You are a front-line worker who has a client in mind.

Here is a sample of the in-house memo you might write to your house director:

To: Adam Lemkow
From: Youssif Amir
Date: October 23, 2009
Re: Enrolling Jaime in the Youth at Work Program

I would like to recommend Jaime for the Youth at Work program. Jaime has been doing a great job of complying with house rules. He has had an outstanding homework completion record this month and has proven he can be organized and focused. He has told me he'd like to start looking for a job, and I think he's ready for employment. He has helped with several small repairs around Phoenix Place and has told me he enjoys the work. Could we put his name forward for a spot in the Youth at Work project?

Youssif's boss has enough information from this memo to make a decision, because he knows the client's case history and has access to any client records he might need to read to make his decision. Imagine that Mr. Lemkow agrees with Youssif and tells him to write to the coordinator of the Youth at Work program to put Jaime's name forward for the program.

Here is a sample of the report, in the form of a letter, that Youssif might write:

Phoenix Place 93 Glen Ridge Ave. Milby, BC, V0N 1L0, 604-555-1232

Jennifer Chow, Coordinator
Youth at Work
Youth Services Programs
87 Red Pine Way
Chetsford, BC V0N 3L8
November 1, 2009

Introduction

I am writing to you on behalf of one of our clients, Jaime Wilson. I would like you to consider Jaime for a spot in your Youth at Work project. I believe Jaime is just the type of candidate for whom you are looking.

History and Mandate of Phoenix Place

Phoenix Place is dedicated to helping male youth who have been in conflict with the law reintegrate into society. It provides residents with a safe and respectful environment in which they can go forward with their lives by completing their high school education through a residential independent study program. It also offers anger management and conflict management programs. Many of Phoenix Place's clients also complete a life skills course in which they learn basic banking and budgeting skills.

Client Profile

Phoenix Place houses low-risk offenders between the ages of 13 and 17. These residents have been convicted of summary offences and are serving their sentences at Phoenix Place. Our typical client has been convicted of executing motor vehicle theft, trespassing, or causing a disturbance. Many of our clients have had drug or alcohol addictions. Those clients must complete drug and alcohol rehabilitation programs as part of their sentencing conditions.

Jaime Wilson's Background

Criminal History

Jaime Wilson is a 17-year-old male who was convicted of motor vehicle theft. This is Jaime's only conviction. Jaime has served five months of his six-month sentence.

Family Support

At the time of his conviction, Jaime lived with his mother, Rosan Wilson. Jaime's mother has provided him with unfaltering support, and she has told us that Jaime will live at home with her at the end of his sentence. Rosan has worked as an office manager at a local refrigeration company for 15 years. No one else lives in the home. Jaime also has a brother, Jon Wilson, living in the city. Jon has been a driver with a local courier firm for three years. Jon also has a small home-renovation business. Jon has visited Jaime often and has said he will continue to support him when Jaime leaves Phoenix Place. Jaime's father, Rick Wilson, was abusive and left the family home when Jaime was 11 years old. Rick Wilson now resides in another province.

Life and Educational Experience

Jaime has no history of drug or alcohol abuse, although he has told us he was drunk when he committed his offence. We enrolled him in a two-week alcohol abuse awareness program, which he successfully completed. Jaime has a grade 10 education. He is currently completing three correspondence grade 11 credits. He works hard and is committed to success. He plans to complete his high school education. Jaime has worked in a local garage and also with his brother on various small construction projects.

Jaime's Progress at Phoenix House

On May 28, Jaime was sentenced and came to Phoenix Place. He kept to himself for the first three weeks. After the third week, Jaime began to interact with staff members and clients. He is well liked by both groups. He is compliant and even-tempered, and wants to have a good future. He began volunteering to help out with minor repairs around the house and has told staff members that he enjoys working with his hands. Jaime completes assigned house chores competently and often offers extra help.

Jaime's Suitability for Your Program

I know Jaime would make a good candidate for your program. He has an aptitude and interest in the construction trade. He has a good work ethic and can be trusted to follow through on whatever is asked of him. When he leaves Phoenix Place he will be living with his mother, and his brother will continue to be a supportive role model.

Jaime has had experience in construction and has expressed his desire to find work in this field. I have talked to him about your program, and he is very excited about the possibility of being chosen. Jaime is remorseful and wants to put his criminal offence behind him. He is the perfect candidate for your program, and I am confident he will be one of your success stories.

Thank you for considering Jaime Wilson for your program. I can be reached at 604-555-7661 if you have any questions, or if you would like to set up a meeting to discuss Jaime's possible enrolment in your program.

Regards,

Youssif Amir

ACTIVITY 3.1 Providing Background Detail

For each of the following scenarios, list three possible background details you would use to make your case. The first scenario has been completed as a sample.

1. You are a front-line worker at an assisted living facility for adults living with developmental delays. You would like the agency that manages the house to have the complete interior of the house painted.
   List background details.

   • *the lounge, which had been the smoking room, still smells of smoke despite having its walls washed several times*

   • *the paint is peeling in all the hallways and in the kitchen*

   • *most bedroom walls have many scuff marks and are hard to clean*

2. You are a part-time worker at a drop-in shelter for women. You would like the house director to let you start a cooking club in which the women would take turns teaching one another how to cook one of their favourite dishes.
   List background details.

   _____

   _____

   _____

   _____

   _____

3. You work at a managed alcohol program. You would like the agency that runs the program to provide sandwiches for the clients. Currently, clients get only pretzels.
   List background details.

   _____

   _____

   _____

   _____

   _____

4. You work in an elementary school as a literacy support worker. You would like the school to buy books that are more relevant to the school's at-risk population.
   List background details.

   _____

   _____

   _____

   _____

   _____

5. You work at a social club for homeless men. You want the agency that funds the club to invest in a TV and a DVD player.
List background details.

_____

_____

_____

_____

_____

3. Use of Persuasion

Good persuasive writing is an art. To get what you want, you must also give readers what they want. How do you do this? To persuade, you must motivate. As discussed in the previous section, you must first give readers all the details they need to make informed decisions. Then you must persuade the readers that they will benefit from granting you your wish. An effective way to do this is to use the "you" focus.

The "You" Focus

This technique requires that you tell your readers that you have their best interests in mind. You can do this by

- Putting the focus on the reader by writing in the second person and using "you" as often as possible
- Telling the readers you know what they want and that helping you will also benefit them

Consider the previous report for the Youth at Work project. The writer's first priority is his client's reintegration into society; however, he knows that the coordinator of the new program will choose candidates carefully, especially at the pilot project stage, because the pilot project must succeed if the program is to continue. The writer must convince Jennifer Chow that Jaime will help make her program a success. He does this by describing Jaime's even temperament, good work ethic, and experience in the field. He also says directly, "He [Jamie] is the perfect candidate for your program, and I am confident he will be one of your success stories."

ACTIVITY 3.2 # Writing Persuasively

Read each scenario below, and think about the proposed ideas. Then list the benefits of agreeing to the proposals.

1. You are a senior correctional officer at a minimum-security institute. Administrators at Corrections Canada have asked correctional officers to recommend outdoor activities for the inmates. You think the inmates should start a vegetable garden. List three benefits that would persuade administrators that this is a good idea.

_____

_____

_____

_____

2. You are a front-line worker at a halfway house for adult male ex-offenders. You want the house director to set up an unused room in the basement as a carpentry workshop for the residents. List three benefits that would persuade the director that this is a good idea.

    _____

    _____

    _____

    _____

    _____

3. You are a child and youth worker at a community centre's after-school drop-in program for teens. You would like teachers at the neighbourhood high school to recommend student tutors to help out at a homework club in the centre. List three benefits that would persuade the teachers that the homework club is a good idea.

    _____

    _____

    _____

    _____

    _____

4. You manage a small apartment building whose residents are women who are fleeing abuse. You would like to organize a running club for the women and need donations of shoes and clothing. List three benefits that would persuade the owner of a local sporting goods store that donating would be a good idea.

    _____

    _____

    _____

    _____

    _____

5. You supervise a support program for ex-offenders attending community college. You want the college administrators to lend you a room where the students in your care could meet to do homework, unwind, and eat snacks. List three benefits that would persuade college administrators to free up a room for you.

    _____

    _____

    _____

    _____

    _____

Chapter Summary

■ Multi-audience reports can be informal, semi-formal, or formal, and can be written in a variety of formats.

■ A multi-audience report must be detailed and complete to allow the reader to make an informed decision.

■ Writers of multi-audience reports must effectively use persuasion to achieve their reports' desired outcomes.

Chapter Exercises

EXERCISE 3.1 Writing Informal Reports

Write informal reports for the five scenarios outlined in Activity 3.2 – Writing Persuasively. You will have to invent all the necessary details to write a complete report.

EXERCISE 3.2 Writing Semi-Formal Reports

Write semi-formal reports for the following scenarios. Remember that semi-formal reports should be 3 to 10 pages long and organized into sections with headings and subheadings. Some details are provided. Please add all the necessary details to write a complete report.

1. You are a full-time front-line worker at a shelter for homeless youth. You would like more funding from your parent agency to start a street outreach program. Here are some details:

   • You will need a van and a budget for blankets, socks, food, and simple medical supplies, such as adhesive bandages and antibiotic cream.
   • You know that many homeless youth are reluctant to visit the shelter. They often get seriously ill with hypothermia, foot ailments, or malnutrition, and must be hospitalized.
   • The city mayor has proposed a no-begging bylaw. If this is passed, homeless youth would be arrested if they begged for food.

2. You work at an agency that provides a variety of services for people living on social assistance. Your supervisor and you have noted an increase in requests for furniture from clients who use your agency. You have decided to write a request to local furniture outlets asking for donations. Here are some details:

   • Your clients are living on social assistance. Many are parents with young children. Most are actively seeking work.
   • The most popular furniture requests are for beds and kitchen tables and chairs.

- You would like the furniture stores to donate furniture. You would also like them to keep the furniture at their warehouses until it is needed. Further, you would like the outlets to deliver the furniture at no cost.
- You are asking each outlet to commit to making three furniture donations and deliveries a year.

3. You are the director of a halfway house for adult male offenders. Your facility is in very bad repair, and you need either funding for a major renovation or a new facility. You have heard that your halfway house's parent agency owns a vacant building that would suit your needs. You have decided to write a report to your supervisor to ask for money to be used for renting the vacant building. Here are some details:

  - The halfway house has a leaky roof and a crumbling foundation. The pipes are made of lead, and city inspectors recently told you the pipes need to be replaced. The furnace is 20 years old. The windows are old and drafty.
  - The vacant building is a three-storey six-bedroom house with a new roof, a large kitchen, rooms on the main floor for an office and a lounge, and a screened-in back porch.
  - The vacant building is on a major bus route and is within walking distance to a YMCA, a convenience store, three banks, and a library.
  - You think the current halfway house building is unsafe for residents.

EXERCISE 3.3 Editing for Unclear Pronoun Use

Rewrite the following passages to fix any unclear pronoun use.

1. I asked each of the clients to fill out the consent form. Oksana and Maraille had questions about the references section. She wanted to know how many references were needed.

2. The correctional officer was escorting the inmate through the gate. Suddenly, for no apparent reason, he hit him.

3. On my way over to my desk, I tripped over the rug and knocked the vase off the desk with my coffee cup and broke it.

4. Neither Nick nor Jimmy admitted to writing the graffiti. They both said they had seen the other with a paint can. Kevin said the graffiti had Ben's tag on it. He said he used to be the best graffiti artist in the city.

5. The paramedic calmed the client down by talking gently to her. She was about 20 years old.

6. The social services worker interviewed the new client. She asked her if she could record the interview. She said yes.

7. Several of the inmates were questioned by the guards about the missing pail and mop. They said that they had last seen the pail and mop at morning head count.

8. The child and youth worker was amazed by how much his client had changed his attitude. He has been drug-free for three months. He is very proud of his progress.

TEMPLATE 3.1

Serious Occurrence Report
Sherman House

Written to
Probation Officer Brendan Forbes, Petersville Probation Office

Written by
Rose LePine, Director, Sherman House

Date of Occurrence
November 18, 2009

Date of Report
November 18, 2009

Occurrence Type
Theft
Assault with a weapon

Client Involved
Resident Andrew Carrea

Additional Individuals Involved
Resident Andrew Carrea
Resident Omar Benzo
Resident Ali Khan
Staff member Lucas Brown
Staff Member Jordan Fergo
House Director Rose LaPine
Police Constable Amy Gallant
Police Constable Albert Dubac

Details of Occurrence

At approximately 10:15 PM on the evening of November 18, 2009, Andrew

Carrea stabbed resident Omar Benzo with a steak knife that Andrew Carrea had

stolen from the residence cutlery cabinet. Omar Benzo suffered serious wounds

to his face and has lost permanent sight in his left eye. Resident Ali Khan and

staff member Jordan Fergo were able to disarm and restrain Andrew Carrea.

Staff member Lucas Brown called 911, and at 10: 35 PM, police constables Amy Gallant and Albert Dubac arrived at the residence. They immediately took Andrew Carrea into custody and left the residence. At 10:38 PM, paramedics arrived at the residence. They examined Omar Benzo and immediately lifted him onto a stretcher and took him to Petersville General Hospital.

Background Details

Andrew Carrea has resided at Sherman House for two weeks. He is completing a heroin-withdrawal program and, until this occurrence, has lived at Sherman House with no behavioural problems. On November 16, Omar Benzo became a resident. On the following morning, when Andrew Carrea learned of Omar Benzo's arrival, Andrew Carrea became agitated and told everyone he came into contact with that Omar Benzo was evil and did not deserve to live. This comment is documented in the log book. Andrew Carrea, however, did not act with aggression or hostility toward Omar Benzo. Instead, staff members noted several times in the log book that Andrew Carrea seemed to go out of his way to avoid Omar Benzo.

After the occurrence, I interviewed resident Lucas Brown who had come forward after the paramedics left to tell me he had information about the relationship between Andrew Carrea and Omar Benzo. Lucas Brown told me that according to Andrew Carrea, Omar Benzo had dated Andrew Carrea's sister and Omar Benzo has beaten up Andrew Carreas' sister badly enough to put her in hospital. Lucas Brown also reported to me that resident Ali Khan told Lucas Brown that Ali Khan saw Andrew Carrea take a steak knife out of the cabinet after dinner the evening of the attack. I then interviewed Ali Khan. According to Ali Khan, the cabinet was left unlocked while kitchen staff member Lucas Brown was preparing the evening meal. Ali Khan was the assigned kitchen helper for that meal. At some point during the dinner preparation, Lucas Brown responded to a knock at the kitchen door, but no one was there. Ali Khan said that the cabinet was unattended long enough for Andrew Carrea, who was also in the kitchen on helper duty, to grab a knife. When Lucas Brown returned from the door, he noticed that the cutlery cabinet door was ajar and locked it. Ali Khan told me he did not see Lucas Brown do a cutlery count before locking the door.

I then interviewed Lucas Brown who confirmed that he had left the cutlery cabinet door open for a brief period, and he also confirmed that he responded to what he thought was a knock at the kitchen door. No one was at the door. In addition, Lucas Brown confirmed that he locked the cabinet right after he returned from the kitchen door. Lucas Brown also confirmed that he did not do a cutlery count before locking the cabinet.

Results of Occurrence

At 10:35 PM on November 18, 2009, Andrew Carrea was taken into custody. He is being held at Petersville Detention Centre pending his arraignment.

Resident Omar Benzo is in Petersville General Hospital.

Lucas Brown has been suspended from duty pending a full investigation of the occurrence.

Rose LePine   *November 18, 2009*

TEMPLATE 3.2

Program Quality Review

East Glen House Resident Carpentry

Program

for

Glengarry County Family and Social Services Advisory

Committee

Margaux Whiteduck

May 30, 2009

Introduction

The East Glen House Resident Carpentry Program has successfully completed its first session. This report provides readers with an overview of the successes and challenges of our inaugural session. As well, this report contains suggestions for program improvements. The implementation of these suggestions will help ensure the continued success of the program. For more information on individual participants, please see the two tables in Appendix A. Table A lists skills taught and participant completion rates. Table B lists participants and their individual projects to date.

Program Overview

This carpentry program operates out of a winterized shed on East Glen House property. The program is voluntary, and six out of eight residents initially signed up. Five have participated regularly; one of the six has dropped out.

Participants attend weekly four-hour classes in which they learn various carpentry skills. Residents are then permitted to spend up to three hours per evening working on individual carpentry projects of their choice. The four-hour class is taught on a volunteer basis by Mr. Bryce Stanley, owner of Stanley's Hardware and Lumber.

The objectives for implementing this carpentry program were threefold:

- Residents would make constructive use of free time.

- Residents would learn skills they could use in their own homes when they are ready to live independently.

- Residents could use the skills to gain employment.

Successes

We are very pleased with this program. Five out of eight residents have consistently attended the carpentry classes. Residents have been able to complete a variety of simple carpentry projects, such as side tables, chests, bookshelves and quilt racks. Many residents have given their projects to friends and family as presents. One resident currently has a contract to make six small display tables for a local bakery. As well, residents have been able to make several repairs to East Glen House. Over the past year, residents have replaced two broken steps, planed and re-hung three doors, added a railing to the back deck, and built a pantry in a kitchen closet.

Staff members at East Glen House have noticed an overall improvement in the morale and camaraderie among residents since the implementation of the carpentry program.

Challenges

We had higher participation than we expected. This meant that residents had to share tools and were often waiting up to 30 minutes for a turn with needed tools. This tool shortage sometimes resulted in raised tempers. The resident who stopped attending informed staff that he couldn't stand the constant waiting, so he didn't want to be part of the program any more.

We were lucky to be able to house the classes and project nights on the premises, but the shed has a leaky roof, painted-shut windows, and poor insulation.

Mr. Stanley was generous with his classroom time, but he wasn't available for consultation during the evening project sessions. Residents were often frustrated when they had to wait for the weekly class to get project questions answered.

Suggestions for Program Improvement

In consultation with program participants, Mr. Stanley, and East Glen staff, we will implement the following changes for improvement:

1. Solicit local hardware and department stores for tool donations.
2. Recruit volunteers who can help residents with their projects.
3. Enlist the residents to help with shed renovations.
4. Increase nightly project hours from three to four.

Conclusion

This carpentry program has been a great success. As noted in the introduction of this report, program participants are very pleased with how the program has improved their lives. We are confident that with a few changes, this program will continue to benefit current and future residents of East Glen House.

When we told the residents we would be writing a report to share the successes of this program, several residents asked if we could invite the advisory committee to East Glen House to view their work. Residents have also volunteered to talk to committee members about how the carpentry program has improved their quality of life, self-esteem, and employment prospects.

It is hoped that this report effectively showcased the East Glen House Resident Carpentry Program's value. Please contact the writer of this report if you have questions or comments. East Glen House staff members would be delighted to have the committee visit our residence to speak with residents and to tour our carpentry shop.

Appendix A

Table A - Skill Taught and Participant Completion Rate

	Weeks 1 - 4 Intro to carpentry	Weeks 5 - 8 Wood Cutting techniques	Weeks 9 - 12 Simple joints	Weeks 13 - 17 Advanced joints	Weeks 18 - 21 Preparing to finish/Choosing the finish	Weeks 22 - 26 Finishing techniques
John P.	✓	✓	✓	✓	✓	✓
Xavier R.	✓	✓	✓	✓	✓	✓
Michael S.	✓	✓		✓	✓	✓
Brett K.	✓	✓	✓	✓	✓	
Riley V.	✓	✓	✓	✓	✓	✓
James H.	✓	✓				

Table B - Resident Participation and Project Completion

John P.	Xavier R.	Michael S.	Brett K.	Riley V.	James H
Small box	Small box	Small box	Small box	Small box	Small box
Plant stand	Plant stand	Bookshelf	Hinged box	Bookshelf	Bookshelf
Quilt rack	Bookshelf	Kitchen utensil rack	CD rack	Serving tray	
CD rack	Corner display shelf	Side table		Small chest	

Short Proposals

After reading this chapter and completing the exercises in it, you will be able to

1. Write a strong summary statement for a program or service proposal.

2. Write a persuasive short proposal that addresses all your reader's questions.

3. Organize your proposal into appropriate sections to effectively highlight your program or service's viability.

Introduction

As a community services worker, you will often see needs within your client group or service areas that are not being met. When this happens, you may have ideas for change that you would like to propose. Proposal writing is common in the community services field. There are many avenues for support for new ideas: the federal government, provincial governments, parent agencies, philanthropists, and members of the public who may have been helped by community services agencies.

Of course, there is competition for the money, facilities, and person-hours required to implement new programs, open new facilities, and expand agency services. You must write strong, persuasive, and organized proposals that turn your ideas into working projects.

SENTENCE POLISHING SKILL: SUBJECT-VERB AGREEMENT

Every complete sentence must contain a subject and a verb. The subject and the verb in the sentence must agree with each other. This means that if the sentence contains a singular subject, the verb must also be singular. Following this rule when writing simple sentences is easy.

Example

That dog is cute.
Singular subject = dog   Singular verb = is

Many dogs are cute.
Plural subject = dogs   Plural verb = are

There are some cases, however, in which deciding whether to use a singular or plural verb can be challenging.

Here are some rules for using subject-verb agreement in your writing:

1. When subjects end in -*thing*, -*one*, or -*body*, they are singular and always take a singular verb.

    ### Examples

    <u>Everything</u> *is* going to be fine.
    <u>Everyone</u> *is* going to the hockey game.
    <u>Everybody</u> *is* delighted to see you.

2. When more than one subject is joined with either/or or neither/nor, the verb will agree with the subject nearest to it.

    ### Example

    <u>Either</u> the dogs <u>or</u> the cat *is* going.

    In the above sentence, both *dogs* and *cat* are subjects. Because *cat* is closest to the verb, the verb is singular. Here is how the rule would work if the writer switched the order of the subjects:

    ### Example

    <u>Either</u> the cat <u>or</u> the dogs *are* going.

    Here is another sentence pair that illustrates how the verb would change depending on which subject the verb followed:

    ### Example

    <u>Neither</u> my sister <u>nor</u> her friends *like* that band.
    <u>Neither</u> my sister's friends <u>nor</u> my sister *likes* that band.

Caution: Present tense single verbs end in *-s* when used with the third-person voice.

Examples

seems (Cheryl seems frightened.)
likes (Robert likes cake.)
takes (The winner takes all.)

3. The subjects *either*, *neither*, and *each* need singular verbs.

Examples

<u>Neither</u> student *has* presented yet.
<u>Each</u> of you *seems* angry.
<u>Either</u> plan *is* suitable.

4. Collective nouns can take either a singular or a plural verb, depending on how the noun is used. A collective noun is a noun that describes a group of people, places, or things:

Examples

Team	Staff	Herd
Jury	Committee	Flock

If the group is acting as a group, use a singular verb:

The <u>jury</u> *is* voting now.

If members of the group are acting individually, use a plural verb:

The <u>team</u> *are* arriving over the next half hour.

5. For distances, time, and amounts of money, use singular verbs.

Examples

<u>Five kilometres</u> *is* too far for me to run.
<u>Five hours</u> *is* a long time to spend in the car.
<u>Five dollars</u> *is* a fair tip.

At the end of this chapter, you will find an exercise to help you practise subject-verb agreement.

Proposal Writing

Writing a proposal may seem daunting, but if you begin with a plan that allows you to break down the work into small steps, you will be more likely to enjoy the task and you will write a sound, persuasive proposal.

Preliminary Writing Steps

Prepare to write your proposal by starting with two steps: writing the summary statement and drafting answers to the five Ws. Completing these two preliminary steps will allow you to focus throughout the proposal writing process.

Summary Statements

A program or services proposal addresses a documented gap or need in the current services provided to clients. A strong proposal starts with a summary statement that clearly and specifically outlines that gap and explains how your proposed program or service will address the problem.

You should address these details:

- Current situation
- Targeted client group
- Proposed service

Here is a sample summary statement:

> Currently, the City of Bramley has one shelter for homeless men. It is overcrowded and at times dangerous, which makes homeless male youth reluctant to use the shelter, even in winter. We propose to open a 60-bed shelter for homeless male youth in the downtown area.

✳ ACTIVITY 4.1 Writing Summary Statements

To complete this exercise you will need to refer to the scenarios described in Exercise 4.1 at the end of this chapter. For each scenario in this exercise, complete the following information.

1. Scenario – Social service worker at a family support centre

    Current situation: _____

    _____

    _____

    _____

    Targeted client group: _____

    Proposed program or service: _____

    _____

    _____

    Summary statement: _____

    _____

    _____

    _____

    _____

2. Scenario – Employment counsellor for adults living with developmental delays

   Current situation: _____
   _____
   _____

   Targeted client group: _____

   Proposed program or service: _____
   _____
   _____

   Summary statement: _____
   _____
   _____
   _____
   _____

3. Scenario – Child and youth worker at an after-school drop-in program at a community centre

   Current situation: _____
   _____
   _____

   Targeted client group: _____

   Proposed program or service: _____
   _____
   _____

   Summary statement: _____
   _____
   _____
   _____
   _____

4. Scenario – Front-line worker at a shelter for women fleeing abuse

   Current situation: _____
   _____
   _____

   Targeted client group: _____

   Proposed program or service: _____
   _____
   _____

Summary statement: _____
_____
_____
_____
_____

5. Scenario – Social services worker at a drop-in centre for low-income seniors

   Current situation: _____
   _____
   _____

   Targeted client group: _____

   Proposed program or service: _____
   _____
   _____

   Summary statement: _____
   _____
   _____
   _____
   _____

The Five Ws

Once you have written your summary statement, you must add your proposal's supporting details. Start the process by mapping out the five Ws, who, what and how, why, where, and when:

Who?

- Who is writing the proposal?
- Who will the proposal serve?
- Who will run the program or service?

What and How?

- What are you proposing?
- What are the particulars of your program or service?
- How will your program or service work?
- How will you measure the success of your program or service?

Why?

- Why are you the right person (or agency) to start this program or service?
- Why offer this program or service at this time?
- Why do your clients need this program or service?

Where?

- Where will your program or service be located?
- Will you need any special facilities?

When?

- How long will your program or service be provided?
- When do you want it to start?

Once you have mapped out the questions, use the information they generate to write your proposal.

Sections of a Proposal

Organize your proposal information into the following sections:

Cover Page

Include the following information on your cover page:

- Name of funding agency (people from whom you are asking money)
- Name of program or service proposed
- Organization/name of proposal writer
- Date

Statement of Purpose (Introduction)

In this section, you provide a brief description of your program or service. Start with your summary statement and then provide more detail.

Evidence-Based Rationale

In this section, you must establish a current need for your program or service. You will provide evidence that proves

- There is currently no program or service such as the one you are proposing; or
- There is a similar program or service in place but it is not adequate to meet the needs of its clients.

You can make use of statistics, letters of support or concern from community members, client testimonials, and letters from expert consultants to provide this evidence.

Description of the Program or Service

Outline the day-to-day workings of your program or service. Answer the following sorts of questions: How often will your program or service run? What will occur in your program? How many staff members will it take to run your program? What will clients do or experience in your program or service?

Measuring Program or Service Success

In this section, describe how you will measure your program or service's success. Will you conduct surveys? Will you interview clients, professionals, or other participating individuals? Will you write client progress reports? Will you track recovery rates?

Staff Profile

You must explain how many staff members it will take to run your program. In this section, you should also include the necessary staff credentials. Provide detail as to what each staff member will do.

Budget

Include a detailed budget for your program or service that shows all costs, such as facilities, equipment, and staff member salaries.

   For a sample of a proposal, please see the appendix to this chapter on page 69.

Chapter Summary

- Writing proposals for programs or services is an integral part of the community and justice services profession.

- Strong proposals get the attention of funding bodies.

- To write strong proposals, answer the five Ws and organize your information into appropriate sections.

Chapter Exercises

EXERCISE 4.1   # Writing Funding Proposals

Using the chapter guidelines and the sample proposal in the appendix to this chapter on page 69, write funding proposals for the following scenarios. You will have to add many details to make your proposals complete. You may write your proposals alone or in pairs. Refer to the work you did in Activity 4.1 to get you started with the summary statements.

1. You are a social service worker working in a family support centre in a small town. At the moment, your town's family support centre helps with housing, food and clothing donations, and emergency counselling. You would like to start a drop-in daycare to provide mothers with a few hours a week of free daycare. The idea is to give the mothers time to run errands without the stress of also attending to their young children. Write a proposal to your funding agency for support and funding to start this program.

2. You are an employment counsellor for adults living with developmental delays. Your job is to find employers who would be interested in hiring your clients. You have an idea for a job fair. You would like to invite 25 local employers to a job fair at which they could meet and briefly interview potential employees. You will need money for a facility rental, hospitality, advertising, and maybe some door prizes. Write a funding proposal to the local Business Owners Association asking for its endorsement and financial support.

3. You are a child and youth worker at an after-school community centre drop-in program in a neighbourhood with a high immigrant population. The program runs sports activities and a homework club. You know that the homework club is not successful. Few students attend and, when they do, they do not accomplish much serious work. You suspect that this is due partly to the tension between the boys and the girls and partly to the lack of tutor consistency. Currently the tutors are local seniors who volunteer when they can. You would like to improve the situation by starting two homework clubs: one for boys and one for girls. As well, you would like to get funding to hire college or university honours students to tutor the youth. Write a proposal to your funding agency to get support and funding for this idea.

4. You are a front-line worker in an east end shelter for women fleeing abuse. Currently, the shelter has no in-house employment counselling for the women. Residents must travel several kilometres to a downtown shelter to access job opportunities through the Internet and get help with resumé writing and job applications. You would like to start a mini-employment centre for your residents. You want the centre to operate out of one of the rooms at the shelter. You would need a few computers, Internet access, office supplies, and the part-time services of an employment counsellor. Your house director supports your plan and has given you the go-ahead to write to your funding agency for support and approval. Write the proposal.

5. You are a social services worker who works with low-income seniors at a drop-in centre. Your job is to run the games room, keep the small library in order, and serve snacks. The seniors appreciate the program. During your time at the centre, you have gotten to know several of the seniors quite well. Many seniors have told you of their musical talents, and many have expressed their regret at not having music in their lives. You have an idea to start a music program. You would like to buy some used instruments (e.g., a piano, banjos, guitars). You would also like to start a choir, which would involve hiring a choir director and getting sheet music. There is a local radio station that runs a golden oldies show every week. Write a funding proposal to the owner of the station asking for financial support for your program.

EXERCISE 4.2 Designing Your Own Program or Service

Think of a program or service you would like to start for the clients with whom you plan to work. Write a funding proposal. Remember to refer to the chapter guidelines and the sample proposal in the appendix to this chapter on page 69.

EXERCISE 4.3 Editing for Subject-Verb Agreement

Circle the correct verb choice from the italicized pairs.

1. Everything in this category *are/is* suitable for your research project.

2. Each *is/are* open.

3. The hospital committee *votes/vote* tomorrow about the new mental health program.

4. Fifty dollars *was/were* all that we recovered.

5. Either his parole officer or his counsellors *think/thinks* he needs protective custody.

6. The 400 kilometres we drove through the winding mountain roads *were/was* nerve-racking.

7. Everyone at the party *has/have* agreed to dance.

8. Neither *are/is* happy about the recent cutbacks.

9. Neither the supervisor nor the workers *know/knows* the new director.

10. The staff *are/is* taking turns visiting the bake sale.

APPENDIX   Sample Support Letters

Student Association
Provincial Community College
3447 Highway 12
Vernon City, SK S0M 6T7

March 23, 2009

Women's Correctional Agency
3300 Broadview Ave.
Sackley, SK S0M 7Y6

Dear Director,

I am writing this letter in support of the Study Support for Success program being proposed by the Amelia Jones Society. I would ask that you please fund this outstanding program.

At Provincial Community College, we celebrate our diverse student body and are proud of our proven record of inclusivity and acceptance. The population that this proposal will serve has been most welcome at the college. I have had the opportunity to talk to some of the teachers who have had these students in their classrooms. They report that the students add valuable life experience to classroom discussions. These students are committed to success and are doing their best to overcome the many challenges they face daily. Teachers and fellow students want to see these young women achieve their goals.

The Student Association will happily donate a classroom for this program and will also help with finding suitable tutors. Please seriously consider funding this initiative. Thank you for taking our support into consideration. You are welcome to contact me at 306-555-8097.

Regards,

A. Dubois

Abigail Dubois

Student Association Director

3-67 Winding Way
Sackley, SK
S0M 2WT

March 26, 2009

Women's Correctional Agency
3300 Broadview Ave.
Sackley, SK
S0M 7Y6

Dear Director,

I am very happy to give my support for the Study Support for Success program. I recently completed the Florist Program at Provincial Community College and was funded by the Ministry of Corrections.

I loved every minute of the program, but I often felt alone and thought about quitting many times. I was able to see the program through with the help of classmates and family. I know some of the other Ministry girls weren't as lucky. I would have loved to have a room set aside just for me and the other women in my circumstance. I think extra tutoring and study skills help and companionship with people who have the same troubles would have really helped those other girls stay at the college.

Please support this program. It's going to be a real help to many people. I would be happy to talk to you about my college experience. You can call me on my cellphone at 306-555-4527.

Yours,

O Joy

Olivia Joy

Letters and Email

LEARNING OBJECTIVES

After reading this chapter and completing the exercises in it, you will be able to

1. Identify the structural elements of business letters.

2. Write persuasive letters for specific audiences and purposes.

3. Differentiate between social email and workplace email.

4. Write workplace email that is informative and free of "texting-style" spelling, emoticons, and errors.

Introduction

You will write hundreds of business letters over your lifetime. You will write them as a consumer, a concerned citizen, and a community services professional. All business letters, no matter what their purposes or audiences may be, follow a standard format. This chapter outlines correct business letter format and explains the process of writing the letter content for a variety of purposes.

Many Canadians use email daily. As a student, you use email to contact teachers, make arrangements for group projects, or contact such school administrative services as the resource centre, the registrar's office, the counselling services department, or the gym. In your personal email correspondence, you use email to keep in touch with friends and family, complete consumer transactions, and research such things as vacation options or professional services. This chapter discusses the value and importance of using professional email conventions for all of your school and workplace electronic correspondence.

SENTENCE POLISHING SKILL: NUMBERS

Knowing when to spell out a number (nine) and when to use the numeric form of a number (9) can be difficult; however, if you take the time to use the correct form, people will respect you as a competent writer who pays attention to detail.

Rules exist as to when you should use each number format. Unfortunately, as is customary with the English language, these rules have exceptions, so be sure to take the exceptions into account in your writing as well. Here are some rules for using numbers in your writing:

1. Spell out the numbers one to nine. Use the numeric form for the numbers 10 and over.

   Example

   I took <u>three</u> clients to the grocery store.

   Exception 1: If you are writing about a mix of numbers below nine and above nine, you can use the numeric form for consistency.

   Example

   It took the 3 clients 20 minutes to get to the grocery store.

   Exception 2: Always spell out numbers that start a sentence.

   Example

   <u>Six</u>teen residents attended the substance abuse workshop.

2. Spell out large round numbers.

   Example

   The ministry has just granted <u>$20 million</u> to mental health care.

3. Spell out ordinals, such as 1st, 2nd, and 3rd.

   Example

   You are the <u>third</u> person to compliment me on my report writing.

4. Here are the correct formats for writing
   - Street numbers: 34 Smith Street
   - Dates: October 24, 2009
   - Simple fractions: three-quarters
   - Percentages: 76 percent
   - Times: 11:15 AM (or 11:15 a.m.)

- Money: $45.00
- Temperature: 23°C

For a more comprehensive list of rules for using numbers, consult your school's resource centre. Most Canadian colleges and universities have online writing labs, which are fantastic student resources.

Letters

All professionals write business letters, and no doubt you have had occasion to write letters for a variety of personal, educational, and professional purposes. No matter what their purpose, all business letters follow the same format. For a sample of a letter of request, please see Template 5.1 on page 89.

Here is a sample of a referral letter:

Susan Gail House
345 Spring Lane
Milby, ON N0D 3R5   ⟶ ①

January 28, 2010   ⟶ 2

Hudson Community Centre   ⟶ 3
17 Winding Way
Milby, ON N0D 4T4

Re: Client referral for Therapeutic Carpentry program   ⟶ 4

Dear Sharyn Barber,   ⟶ 5

I am a child and youth worker at Susan Gail House and am writing this letter to ask that you enroll my client Frankie Peal in your Therapeutic Carpentry program. Frankie has been a client at Susan Gail House for six months. He has successfully completed our drug rehabilitation program and is looking forward to becoming independent. I believe Frankie would be a perfect candidate for your program.   ⟶ 6

Frankie is a 17-year-old youth with a six-year history of addiction. He has been very successful at Susan Gail house, and he is now ready to live on his own. He has signed a lease for a bachelor apartment in a subsidized housing complex and has a job in the warehouses of Cuddles pet stores. Frankie has a positive attitude and a desire to turn his life around. Frankie is concerned, however, that he will have too much time on his hands once he leaves the house, and he has asked me to help him find a hobby or pastime. Your Therapeutic Carpentry program would be perfect for Frankie. He is good with his hands and takes pride in a job well done. He is a respectful young man, and he would be a keen student in your program.

Thank you for considering Frankie for your Therapeutic Carpentry program. I know he will succeed, and he would be a positive addition to your class. If you would like to discuss Frankie's suitability, you can reach me at 513-555-9986.

Regards, ————————————————→

7

Brittni Marie ————————————————→
Child and Youth Worker

Encl.: ————————————————→ 8

Structure of a Business Letter

The sample referral letter has a number beside each of the elements that you must include in a letter. The following list provides information about these elements.

1. Writer's Address

The writer's address appears first on the page. If you are using letterhead, the address will already be on the page and may be centred, left justified, or right justified. If you are not using letterhead, write your address along the left margin. Your name should *not* appear in your address block.

2. Date

The date should be written with the month spelled out, and the day and the year should follow in numerals.

3. Reader's Address

The reader's address should begin with the reader's name and title. Follow this with the reader's complete mailing address, including a room or suite number if applicable.

4. Subject Line

You can include a subject line in your letter that tells the reader to what your letter refers. A subject line is especially useful when you are writing a job cover letter in response to a particular job competition number. It is also useful when you are writing to someone who has more than one role or project, and you want to draw your reader's attention to a specific topic.

The subject line is optional.

5. Salutation

Here are samples of the ways you can write your salutation:

- Dear Professor Smith
- Dear Mr. Smith
- Dear Mr. Frank Smith
- Dear Professor Frank Smith

Your goal is to get your reader's attention and show respect. Use the salutation you think is suitable for your purpose. *Never* begin your letter with "To Whom It May Concern." Writing this text tells the reader that you could not be bothered to find out his or her name. Make the effort to learn your reader's name (including the correct spelling). If this is impossible, you can use a generic job title (e.g., Dear Recruiting Officer, Dear House Director, Dear Human Resources Manager).

6. Body of Letter

Letters should be only one page long if possible, and you should aim to write three to five body paragraphs. Using three paragraphs is the most common set-up, but sometimes you need to organize your letter into more paragraphs. For more information about what information goes into each body paragraph, see "Body Paragraphs" below.

7. Complimentary Closing

Use a closing that you feel is appropriate (e.g., Regards, Yours truly, Respectfully, Respectfully yours, Thank you). There is no rule that dictates which closing you should use in which circumstance, but as you can see from these choices, some closings sound more formal than others.

If you use a two-word closing, do not capitalize the second word. This is a small detail, but paying attention to the small details is often critical to your success.

8. Enclosures

Sometimes you send additional information along with your letter. This may be a resumé, an invoice, or extra information you want the reader to have. Let the reader know you have included this information by writing "Encl.:" at the bottom of your letter. If you forget to include the enclosure, your reader will know to ask you about it.

Body Paragraphs

Most of the letters you will write in the community services field will ask for someone's help. Your letters need to be complete, professional, and persuasive so the reader will be well informed and will want to help you.

Here are some scenarios for which you might write a letter:

Requests for information or help

- For services for a client
- For funding
- For information (e.g., housing, education, employment) on behalf of a client
- For guest-speaker appearances

Referrals for clients

- Character referrals
- Program referrals

- Court referrals
- Housing referrals

Each of the body paragraphs of the letter has a specific purpose.

Opening Paragraph

Your opening paragraph will be brief. Its purpose is to introduce yourself and tell the reader why you are writing. Specifically state what you want your reader to do or to know.

Middle Paragraph(s)

A letter is brief, so in the middle part of a letter you must succinctly give the reader all the pertinent facts he or she needs to help you. For example, if you are looking for funding, you need to tell the reader how much money you would like and provide exact details about how the money would be spent. If you are referring a client for a service (as in the scenario used by the sample letter earlier in this chapter), you would need to tell the reader specifically how your client would benefit from the service and how your client is a good fit for that service. If you are responding to a complaint, your middle paragraph(s) would address each of the complainant's concerns.

Typically, you should try to write one middle paragraph. This is not always possible however, so write more than one middle paragraph if necessary. Remember that an ideal letter is just one page long.

Closing Paragraph

In this paragraph, you must do three things: remind the reader what your message is, thank the reader, and provide your contact information.

Audience-Specific Writing

Your letters should have a reader focus. You can do this by using the word "you" in your letters and by telling the reader specifically what he or she will gain by helping you. This technique is also known as using the "you" focus. The "you" focus tells the reader you have his or her interest in mind, and this should make him or her more receptive to your request.

Example of text using a writer focus:

> I would like to further my career in social services by completing my placement at Elm Street Shelter.

Example of text using a reader focus:

> My ability to anticipate conflict and mediate effectively will make me a real asset to your shelter team.

Formatting

Professionals commonly write letters in full-block format. This means that the writer does not indent the lines at the beginning of paragraphs and lines up the parts of the letter (e.g., addresses, dates, opening, closing) along the left margin.

The sample letter on page 73 is in full-block format. This style of letter writing became popular with the widespread use of computers. If you prefer the look of indenting the first line of each paragraph, however, it is perfectly acceptable to do so.

Letter templates are available on most word processing programs. The templates are easy to use, so if you are new to letter writing, try them out.

ACTIVITY 5.1 Brainstorming to Write a Letter

Imagine that you are a probation officer with a client who has finished probation, lined up a job, and found an apartment she would like to rent. The landlord requires a letter of reference from you. Put yourself in the landlord's place to decide what information the letter should contain to convince the landlord to accept the client as a tenant.

1. Can you think of three questions a landlord might have about the client? Write these questions in the spaces provided below.

   i. _____

   _____

   ii. _____

   _____

   iii. _____

   _____

2. After you decide what information you need to include, draft a quick outline that details what information will go in each paragraph in the space below.

   _____
   _____
   _____
   _____
   _____
   _____
   _____
   _____

Here is a sample outline that you might have written for the above scenario:

Paragraph one

- *Make personal introduction.*
- *Introduce client.*
- *Request that landlord rent an apartment to client.*

Paragraph two

- *Give the landlord a brief history of client.*
- *Discuss client's financial stability.*
- *Discuss client's lifestyle.*
- *Discuss client's commitment to succeed.*

Paragraph three

- *Thank the landlord for considering the request.*
- *Provide contact information.*

Here is a sample of the completed letter:

Riverside Probation Office
77 Front St.
Riverside, NL A0K 1P8

January 19, 2010

Mr. Steve Wilson
Southview Apartments
578 East Druer St.
Riverside, NL A0K 3T9

Dear Mr. Steve Wilson,

I am a probation officer with the Riverside Probation Office. I am happy to write this character reference letter on behalf of my client Marissa Mikula. Please allow Marissa to become a tenant in your building. I have known her for eight months and am confident that she will be a responsible tenant.

Marissa has a full-time job with Sandra's Cleaning Service and leads a very quiet life. Marissa has told me she disclosed her criminal past to you. I can confirm that Marissa has no known history of drug abuse or violence. She does not socialize with persons in conflict with the law and is trying very hard to put her past mistakes behind her. Marissa is determined to lead a good and productive life.

Thank you for considering Marissa as a tenant. I know she would be responsible, clean, and timely with her rent payments. If you would like to discuss Marissa's suitability with me, you can reach me at 709-555-7761.

Regards,

Myrna Gallant
Probation Officer
Riverside Probation Office

ACTIVITY 5.2 Writing Letter Outlines

Read the following scenarios and decide what information should be included in each paragraph of each letter. The first scenario outline is completed as an example. Remember to consider what information the reader would need to respond favourably to the letter.

1. A correctional officer in Alberta is writing for information about employment opportunities at Ontario's Marksville Penitentiary. His wife has been transferred to a job in Ontario.

Paragraph one

- *Make personal introduction.*
- *Make employment information request.*
- *Explain the interest in moving to another province.*

Paragraph two

- *Discuss experience as a correctional officer in Alberta.*
- *Talk about any additional completed training.*
- *Discuss any promotions, commendations, and awards.*

Paragraph three

- *Thank the reader for her help.*
- *Include contact information.*

2. A front-line worker in an open-custody halfway house is writing to a youth wilderness challenge program. He wants to bring his youth to the camp for a one-day adventure at a reduced rate.

Paragraph one

_____

_____

_____

Paragraph two

_____

_____

_____

_____

Paragraph three

_____

_____

3. A director of a new homeless shelter is writing to a local grocery store to request discount prices for the home's kitchen.

Paragraph one

_____
_____
_____

Paragraph two

_____
_____
_____
_____

Paragraph three

_____
_____
_____

4. A child and youth worker is writing a letter to the aunt and uncle of a child in care. The child has asked if her aunt and uncle could visit her.

Paragraph one

_____
_____
_____

Paragraph two

_____
_____
_____
_____

Paragraph three

_____
_____
_____

5. A front-line worker in an emergency shelter for young women is writing a letter to the high school in a young woman's hometown. The worker wants to get the young woman's school records forwarded to the high school at which the young woman is registering for the fall.

Paragraph one

_____
_____
_____

Paragraph two

_____

_____

_____

_____

_____

Paragraph three

_____

_____

_____

ACTIVITY 5.3 ## Writing Letters

Use the scenarios above to write complete business letters.

Emails

Electronic mail has become the most common form of written communication in the work-place. Before email, if you wanted to contact someone outside your organization, you would write them a letter (formal) or call them on the telephone (informal). If you wanted to contact someone within your organization, you would call them or send them a memorandum (or "memo"). Organizations still use memos, but most workplaces use email more frequently. For more information on writing memos, please see "Memoranda" on page 84.

Because email is so quick to write, and because we use it so frequently, we tend not to follow formal writing conventions when we use it. This is perfectly acceptable when writing to friends and family; however, as illustrated in the following two messages, writing that is too informal, littered with errors, and missing information is not acceptable in the work-place and will likely achieve the opposite of what you want.

Here is a sample email from a student:

Subject: late!!!!!

hey prof Smith,

I'm going to be late handing in my report. i am having a problem with my research :(can you give me an extension.

Thanks :) :)

What are the problems with this email?

_____

_____

_____

_____

_____

Here is another email to the same professor. To which student is the professor more likely to respond positively?

Subject: Extension request for ENL1866 report

Good morning, Professor Smith,

Unfortunately, I will not be able to hand my report in on time. I am having some problems with my research. Can I please have an extension until Tuesday?

Thank you for considering my request.

P. Sanchez
ENL1866, section 040

Note: Poor writing, no matter how simple the task it conveys, undermines your professionalism and weakens your credibility.

Here are five rules for using email as a professional writing tool:

1. Begin with an appropriate salutation. Address your readers using their correct names and titles, if appropriate. If you are writing to a teacher, you might use "Dear Professor"; if you are writing to a work colleague, use the person's name.
2. Use a clear, informative subject line. It is your job to make sure your intended reader opens your email.
3. Include all necessary information to allow your reader to consider your request (e.g., dates, times, costs, actions required).
4. Do not use emoticons (e.g., ☺, :>), "texting-style" spelling (e.g., lol, thx, fyi), or a lower case "i" as a personal pronoun.
5. End with your complete name and any other relevant identifying information. This could include the number of the course that you are taking, the job title you have at work, or the invoice number you are referencing for a service call.

Note: Do you remember the "you" focus mentioned in the letter-writing section of this chapter (see page 76)? Using the "you" focus is also important for email.

As with any other writing, edit your email before sending it. No matter how brief your email might be, you should still take the time to ensure it contains no grammatical, spelling, or sentence structure errors. In your field, every piece of writing you do becomes part of a client's record, and is, thus, a legal document that a variety of professionals may see.

ACTIVITY 5.4 Editing Poor Emails

Edit the following emails for errors and missing information. Remember to consider tone and language (for more information, see Chapter 1) and refer to the five rules for writing emails on page 82.

Here is an email from one colleague at a shelter for youth to another:

> Subject: You're bad report
>
> Hello Nancy,
>
> Do you have time this week to meet? Here's why. You screwed up Tarek's report and ya better fix it before boss lady gets back from the conference ☺
>
> H

_____

_____

_____

_____

_____

_____

Here is an email from a front-line worker in an adult male open-custody halfway house to a program director from Correctional Services Canada:

> Subject: Referral for Bea Smithens
>
> Good morning, Jill,
>
> I need you to let my client into your anger management program right away. His name is Mohammed Hartley, and he's in need of your program. Let me know asap.
>
> Harry Chin

_____

_____

_____

_____

_____

_____

_____

_____

Here is an email from a social service worker requesting some housing information for a client:

Ms. Lu,

Please send me some information about how to get my client onto a waiting list for subsidized housing. The sooner the better would be good, as my client is currently living in a shelter and is very unhappy there.

I appreciate your help ☺

Rosalind Ward

_____

_____

_____

_____

_____

_____

_____

_____

Memoranda

As noted earlier in this chapter, today email often replaces the memo as a communication tool; however, most large organizations use both the memo and email as communication tools among employees. For samples of memos, please see the memo below and Template 5.2 on page 90.

Here is a sample of a memo:

Memorandum

To: Kerri Monahan ——————————————————► 1

From: Shannon MacDonald ——————————————————► 2

Date: January 12, 2010 ——————————————————► 3

Re: Conflict Resolution Training ——————————————————► 4

I would like to take a conflict resolution course offered through the college next month. It is an eight-hour course that costs $250.00. Upon completion of the course, I will have a Level I Certificate in Conflict Resolution. The course is offered in the evening, so I would not need any time off work. At our staff meeting last month, you asked us to think about professional development that would benefit both us and the house. I think this course would be worthwhile. Can we meet to discuss this?

K.M.

As with letters and email, your memos should be professional and complete. Remember to give your reader enough information to allow her to say yes to your request.

Memorandum Headings

The sample memo has a number beside each of the elements in the memo's heading. The following list provides information about these elements.

1. To: Put the person's full name here. You can also include a job title, if appropriate. Even if you know the person to whom you are writing very well, your memo is a workplace document and should reflect the same high standards you employ for the rest of your professional writing.
2. From: Again, put your complete name here. Memos will not necessarily be read only by your intended reader. Your memo should stand alone as a workplace document. If a new employee takes over a file, she must be able to identify the writers of all the documents in the file.
3. Date: Use the date on which you are writing the memo.
4. Re: This is your subject line. Write four or five words that explain the contents of your memo.

Note: You may put your initials at the end of a memo, but memos do not have a salutation or closing.

Chapter Summary

- Professionals write letters to people outside their organizations.

- Professionals write emails to people outside or inside their organizations.

- Professionals write memos to people within their organizations.

- Professionals write letters, emails, and memos in standard formats regardless of their purposes or intended audiences.

- Letters, emails, and memos are professional documents that must be written with care.

Chapter Exercises

EXERCISE 5.1 Writing Emails and Memos

Use the following scenarios to write emails. Each scenario needs more information to complete the task, so remember to add all the necessary extra details that will allow the reader to make a decision.

1. You are a front-line worker in a halfway house for female offenders. Write an email to your supervisor asking for a day off one day next week.

2. You are a social services worker at a youth group home. One of your residents is having anger management issues at school, and you would like to set up a meeting with her high-school guidance counsellor to discuss a behavioural plan for her. Write an email to the guidance counsellor asking for a one-hour meeting. The guidance counsellor is new to the school, so this will be your first correspondence with him.

3. You work in an open-custody halfway house for male offenders. You would like to take two of your residents on an outing to a Junior A hockey game. You have permission from your supervisor, but you would like to invite another worker along to help you out. Write an email to your colleague asking if he would like to join you.

4. You are a front-line worker in a home for adults recovering from substance abuse. You would like to start a therapeutic cooking program for the residents. Write a memo to your house director asking for a meeting to discuss the program.

5. You are a social services worker at an agency for the homeless. You work in a rural branch of the agency. You want some advice about how to help a client who would like to relocate to the city and find work there. Write an email to a colleague at the city branch asking for her help.

EXERCISE 5.2 Writing Letters ✳

Use the following scenarios to write letters. Each scenario needs more information to complete the task, so remember to add the necessary details that will allow the reader to make a decision.

1. You are a front-line worker in a halfway house for youth. Four neighbours have gotten together to write a letter of complaint about your residents. It seems that the youth have been hanging out in front of the house in the afternoons and have been harassing the people coming home from work by making inappropriate comments and using foul language. The letter writers have threatened to take their complaint to their MPP if the problem is not dealt with to their satisfaction. Write a letter to the neighbours in response to their complaint.

2. You are a front-line worker in a closed-custody facility for young offenders. You are planning a speaker's series called "Pursuing Your Goals" for the youth. Write a letter to a local well-known athlete asking him to do a one-hour presentation to the youth.

3. You are a child and youth worker at an agency for homeless youth. Write a letter to a local veterinary hospital requesting that a veterinarian volunteer twice a year to hold a clinic for the youths' dogs.

4. You are a child and youth worker at a youth services bureau. One of your clients needs a reference letter for his new employer. The employer has requested a character letter that confirms the client has housing and is of good character. Write a referral letter that confirms the client's readiness to be a responsible employee.

5. You are a social services worker at an outreach program for adults recovering from substance abuse. You would like the program director to consider one of your clients for a spot in a new therapeutic art class the local hospital is offering. Write a letter to the director of the program in which you discuss your client's suitability for the program. Also, discuss how this program will be of specific benefit to your client.

EXERCISE 5.3 Using Numbers

Identify and correct the errors in number usage in the following sentences. Some sentences may be correct.

1. 73 people graduated in my child and youth worker program this year.

2. I like to think I am right 90% of the time.

3. A new mental health facility for adults has opened at fifty-seven Byron St.

4. I was able to save $35.00 on my gym membership by signing up for an orientation session.

5. It's very hot out. I read 32 degrees Celsius on my thermometer when I left the house at two P.M.

6. There are two forks and 16 spoons missing from the cutlery drawer.

7. I have used one-third of my yearly sick leave.

8. There were 2 clients who went missing yesterday.

9. The odds are that I will have written 50 percent of the essay by tomorrow afternoon.

10. Why don't we take all eleven clients to the baseball game?

TEMPLATE 5.1   Sample Letter of Request

North Bankroft Family Services
67 10th Street East
North Bankroft, SK S4K 4R4

March 4, 2009

Owen Markham
Plainsview County Equestrian Association
Box 320
County Road 30
Plainsview, SK S4T 3W4   → subjective?

Dear Mr. Markham,

I am a child and youth worker with North Bankroft Family Services. I am writing to you for information about your therapeutic riding program for youth. I often have clients who I think would be good candidates for your program, but I don't know very much about how your program works. Could you please take some time to answer the following questions?

What exactly is a therapeutic riding program?
Who are your typical clients?
Who is eligible for subsidized lessons?
Do youth need previous riding experience?
Do you have a waiting list?
What type of application or referral process is required for consideration of acceptance?

I appreciate your help. My email address is mccurdyf@bfs.ca, but I can also be reached by telephone at 306-555-0908 if you prefer to answer my questions with a phone call. I look forward to hearing from you.

Regards,

Fiona McCurdy

Fiona McCurdy

TEMPLATE 5.2   Sample Memorandum

Memorandum

To: All Staff of the Bay Street Shelter
From: Don Houston
Date: May 1, 2009
Re: Congratulations to all of you!

Our shelter has just won the Eddy Smythe Human Services Award. As you may know, this award was created in honour of Eddy Smythe, a street youth who died from a drug overdose five years ago. Eddy's parents initiated the award four years ago as a thank you to all the people working to keep street youth safe.

Specifically, the award is given to a human services agency that best models caring and respect for the clients it serves. Our colleagues at the Jayne Street Centre nominated us for this award.

Please join me on Friday afternoon at 3 o'clock in the front lobby. At this time, Mr. and Mrs. Smythe will be presenting the award. Refreshments will be served. As you see our clients, please tell them that we would like them to attend.

Well done everyone!

Brochures and Newsletters

LEARNING OBJECTIVES

After reading this chapter and completing the exercises in it, you will be able to

1. Produce informative and well-written brochures and newsletters.

2. Choose the appropriate language to deliver a succinct and informative message.

3. Apply basic design principles to create visually appealing brochures and newsletters.

Introduction

Brochures (also known as pamphlets) are a common form of communication. You have seen brochures in the form of takeout menus delivered to your door, study skills brochures available at school, health advisory pamphlets available at your doctor's office, and advertisement notices of upcoming community events at your library or community centre. Brochures are a popular method of distributing information to a large audience; they are inexpensive to produce, easy to distribute, and, if written and designed well, an effective way to circulate your message. People in the community services field use brochures for many purposes: to advertise programs (e.g., conflict management), introduce a new service (e.g., dog care for homeless individuals with pets), and provide clients with tips (e.g., job search, healthy eating, or basic banking techniques).

Newsletters are published with specific audiences in mind. They can be in letter or newspaper format. Newsletters are usually printed on 8½ × 11" paper. They can also be printed on 17 × 11" paper and folded in half to form booklets. Employee newsletters can contain such items as changes in policy, news of promotions, success stories, and funny anecdotes. Newsletters written for clients (sometimes with the help of clients) can contain success stories, creative writing, and tips about such things as healthy eating, shopping bargains, and social clubs. Newsletters can also be written for the community to inform community members about interesting or exciting events at your centre, shelter, or facility. A community newsletter helps build good relations with the community and reminds people of the good work you do.

SENTENCE POLISHING SKILL: CAPITALIZATION

Rules for capitalization are not complicated, but some people tend not to use correct capitalization, especially in emails because of their casual nature. Ignoring capitalization rules may be acceptable in personal communication, but you must use proper capitalization in workplace correspondence. Here are some rules for capitalization.

1. Capitalize all proper nouns (people names, place names), including the pronoun "I."

     Examples

     Sally
     Ms. Yohansen
     Manitoba
     Montreal
     Pacific Ocean
     Edina Street

2. Capitalize the first word in a sentence.

3. Do not capitalize a job title unless it is followed by the name of the person who holds that title.

     Examples

     I was seen by the doctor.
     I was seen by Doctor Luengos-Santos.

     She was nominated for the scholarship by the professor.
     She was nominated for the scholarship by Professor Rissler.

     I went to Halifax to visit my aunt.
     I went to Halifax to visit Aunt Wendy.

     Exception – Do capitalize titles of great importance.

Examples

Prime Minister
President
Pope
Queen
Buddha

4. For titles of books, films, plays, etc., capitalize all words except articles (e.g., the, an, a) and prepositions (e.g., of, on, as, at); however, capitalize articles and prepositions if they begin the title.

Examples

The Karate Kid
Buffy the Vampire Slayer
The Art of Motorcycle Maintenance
The Good, the Bad, and the Ugly

5. Use lower-case letters for seasons and directional cues unless they are part of a place name or in a title.

spring	north
summer	south
fall	east
winter	west

Examples

I can't wait for spring.
I can't wait for Spring Break.

I am going south this summer.
He has visited the South Pole.

At the end of this chapter, you will find an exercise to help you practise using capitals.

Brochures

Typically, brochures are written on two sides of an 8½ × 11" paper and each side is divided into three small pages, or panes. For a sample of a complete brochure, please see Template 6.1 on page 106.

Side A

Side B

When folded, a six-pane brochure is created, with each pane being approximately 8½ × 3½". Most word processing packages now come with a simple-to-use brochure template. The template instructions include guidelines for putting the correct information in the correct pane so that the brochure reads as it should.

You can also design a simpler brochure by folding an 8½ × 11" paper in half to create a four-pane brochure.

Brochure Contents Positioning

Typically, brochures in the community services field are written to announce new programs or services or to give clients an informational handout. Here is one way you could order the information in a program or services brochure:

Side A

Program description	Program details	Benefits of program

Side B

Particulars (times, dates)	Contact/registration information	Cover page (program title, graphic or logo if applicable)

ACTIVITY 6.1   Organizing a Brochure

Setting up a brochure properly can be challenging. Try this quick exercise to understand how a brochure works. This activity will be particularly useful for visual learners.

1. Take an 8½ × 11" paper, and label one side "Side A" and the other side "Side B."

2. Fold the paper into thirds. Open up the paper again.

3. Using the brochure template above, record the information onto the paper you folded.

4. Fold the paper again. You should now have a brochure template for future reference. Although you will create your brochures electronically, this mock-up will be a useful reminder for you.

Design Considerations

Brochures are meant to catch the reader's attention and be read quickly. The following design considerations will help you create brochures that your audience will take the time to read.

The trick to designing a good brochure is to use variety while also keeping your brochure simple. Too much text, too many colours, an inappropriate typeface choice, or too many graphics will turn your brochure into an unorganized, hard-to-read document.

Effective Use of White Space

Remember, white space is the space on the page that contains no text or images. Using white space effectively draws the reader's attention to the text. Your readers can gather information quickly and skip to the content in which they are most interested.

Consider this first pane of a brochure below:

Healthy Eating on a Budget Saturday Workshop

Everyone knows the importance of healthy eating. Healthy eating is important for increasing brain activity, developing bones and muscles, boosting energy levels, strengthening the immune system, and fighting off serious disease. Healthy eating can also reduce your risk of getting serious diseases, such as cancer. Everyone would like to eat the right foods; however, this is often impossible when you are on a tight budget. Good nutrition requires planning and creativity when you have a small food budget. We would like to help you make healthy, affordable food choices that work with your lifestyle and eating preferences. This six-hour Saturday workshop will give you valuable tips for bargain shopping using the Canadian Food Guide. We will also give you tasty recipes to take home and will end the session with a hot meal.

The page includes some valuable information, but your intended audience will likely have stopped reading before they got to the workshop details. They may not even realize that this is a workshop opportunity.

Here is a different version of the pane:

Healthy Eating on a Budget
Saturday Workshop

Benefits

- increased brain activity
- well-developed bones and muscles
- boosted energy levels
- strong immune system

On a tight food budget?

Come to our six-hour Saturday workshop to learn how you can stay in your budget and still eat healthily.

You will get

- tips for bargain shopping using Canada's Food Guide
- simple and tasty recipes
- a hot meal at the end of the session

Source: Shutterstock

Use of Bullets

When you can, put your information into bulleted lists. This makes your information stand out and faster to read. The faster your readers can absorb the information you have to offer, the more likely they will be positively encouraged to take action.

Here is a descriptive paragraph about an anger management seminar:

Our anger management seminar will help you identify your own emotional triggers and develop techniques to manage your negative emotions. You will also learn how to recognize escalating emotions in others and some negotiation skills that will help avoid potential conflict.

Here is the same information in a bulleted list:

Our anger management seminar will help you

- Identify your emotional triggers
- Manage your negative emotions
- Recognize escalating emotions in others
- Use negotiation skills to avoid conflict

Remember to use parallelism for the items in your list. For more information about parallelism, see "Sentence Polishing Skill: Parallelism" on page 2.

ACTIVITY 6.2 Using Bullets

Change these informative paragraphs into bulleted lists.

1. This three-hour vegetable cooking class will teach you the basics of making healthy vegetable dishes that you can serve at suppertime. You will learn which vegetables have the highest nutritional content. You will learn a variety of vegetable cooking methods. You will also learn about some popular vegetables from a variety of countries.

   In this three-hour vegetable cooking class, you will learn

   - _____
   - _____
   - _____
   - _____

2. This is a one-day study skills workshop offered to high-school students who are following alternate high school delivery programs or who are reintegrating back into the high school system. In this class, you will learn how to stay focused and take effective notes, study for tests and exams, and write tests and exams.

   At the end of this one-day study skills workshop, you will be able to

   - _____
   - _____
   - _____

3. This two-hour banking seminar is for people new to banking. In this seminar, participants will learn how to open and use chequing and savings accounts, how to use online banking for bill paying, how to use a bank debit card, and how to establish a good credit rating.

   This three-hour banking seminar is for people new to banking. During the seminar, you will learn how to

   • _____
   • _____
   • _____
   • _____

4. This two-day weekend course will introduce you to the basics of badminton. You will learn the rules of the game, perform a variety of practice drills, and play three games. You will end the weekend by participating in a round-robin tournament.

   During this two-day introduction to badminton, you will

   • _____
   • _____
   • _____
   • _____

Use of Images

Images add visual appeal to your brochure. You can use clip art, photographs, drawings, maps, or charts.

Here are some tips for using images effectively:

1. Choose images that relate directly to your brochure's theme. If you are writing a brochure about healthy eating, do not include images of flowers. The flowers may look nice, but they have nothing to do with healthy eating.
2. Make sure your image is clear when reproduced. You will need to print out a sample brochure to determine the image's quality.
3. Size your image appropriately. For example, if you are including a map with your brochure, make sure the writing on it is large enough to read.
4. Credit your sources. Depending on where you get your image, you may need to reference its source. For example, if you use a pie chart that you got from a website, your must credit that website.

Use of Colour

If you are fortunate enough to work at an agency or facility that has a colour printer that you can use for large print jobs, you can add colour to your brochure.

Here are some guidelines for using colour effectively:

1. Choose an ink colour that is easily read.
2. Keep all the text in one colour, and use boldface or italics to denote headings.
3. If you cannot use a colour printer, consider using coloured paper. Choose a pale coloured paper that will not make your text hard to read.

Typeface Selection

A unique typeface can add visual appeal to your brochure; however, as with any design choice, you must be careful when choosing a typeface. For example, this brochure might annoy or turn off its intended audience: it is hard to read and the typeface style does not suit the brochure's content.

Healthy Eating!

Benefits

- increased brain activity
- well-developed bones and muscles
- boosted energy levels
- strong immune system

On a tight food budget?

Come to our six-hour Saturday workshop to learn how you can stay in your budget and still eat healthily.

You will get

- tips for bargain shopping using Canada's Food Guide
- simple and tasty recipes
- a hot meal at the end of the session

Here are some tips for choosing effective typefaces:

1. Choose a typeface that is easy to read.
2. Choose a typeface that is appropriate for your task.
3. Do not use too many typefaces. Keep your brochure simple.
4. Do not capitalize all of the text.

Content Considerations

You can use a variety of techniques to make your brochures informative, appealing, and easy to read.

Use of Outcome-Based Words

Be specific about what the reader will gain. Use such outcome-based words as

• Understand	• Practise	• Change
• Learn	• Use	• Recognize
• Perform	• Organize	• Succeed

Use of the "You" Focus

Using the "you" focus allows your reader to feel a personal connection to the service, program, or tips offered in the brochure.

Here is a sample that does not use the "you" focus:

Job Search Skills Workshop

Participants in this two-day workshop will learn how to

- write a strong, persuasive cover letter
- produce an informative and organized professional resume
- dress appropriately for an interview
- speak with confidence at the job interview

Here is a sample that uses the "you" focus:

Job Search Skills Workshop

In this two-day workshop, you will learn how to

- write a strong, persuasive cover letter
- produce an informative and organized professional resume
- dress appropriately for your interview
- speak with confidence at your job interview

Use of Questions

Using questions is an effective way to reach your intended audience. If your readers want answers to the questions you include, your pamphlet will be of great value.

1. Use questions as attention grabbers.

   Examples

   How long have you struggled with anger?
   Do you want to improve your physical health?

2. Use FAQs, or frequently asked questions.

ACTIVITY 6.3   # Evaluating Brochures

Find brochures from places you visit (e.g., the grocery store, library, community centre, school), and bring them to class. Get into groups of four, and share the brochures you have found. Critique the brochures using the design considerations and content discussed in this chapter. Then rank the brochures you found from best to worst. Share your findings with the class.

Newsletters

Newsletters vary in length. Some organizations send out a one-page newsletter once a week. Other organizations send out longer publications less frequently. Newsletters can also be distributed electronically. The content of newsletters is usually provided by more than one person. For samples of newsletters, please see the newsletters in this section and Template 6.2 on page 108.

Letter-Style Newsletters

Letter-style newsletters are usually one page. People often send this type of newsletter to family and friends at the holidays along with a holiday card. Professionals such as doctors, dentists, and realtors use them to let their clients know of changes and improvements in their services. You might send these to your clients' families or to other agencies who refer clients to you.

Here is a sample newsletter that uses the letter style:

House of Peace – April Newsletter

Spring Has Sprung!

We know you've all been enjoying the warm spring weather, and you'll be happy to hear that we will be putting our patio furniture out this weekend. If anyone has some free time on Saturday morning at 10:00, you can meet us on the patio and help out. We'll serve a snack afterward.

Our garden club will be starting up soon. Please put your name on the list in the lobby if you would like to join the club. No experience necessary—just come and play in the dirt! Once again, we will be renting a van to take a trip to Rose's Nursery for new flowers. We could use your help choosing from all the wonderful plants at Rose's.

Does anyone know what happened to the brown snow shovel that used to be in the front lobby? It has been missing since the last storm in March. If you used it to clear your walkway, please put it back in the front lobby. Thanks!

Mike from apartment D is starting a checkers club. Game night will be every Tuesday from 7:00 to 8:30 in the basement lounge. Mike would like you to visit him at his apartment if you want to join the club. The games start this Tuesday.

Don't forget to put your clocks ahead one hour next Saturday!

Happy spring!

From your management team, Hermes, Kathy, Frank, and Nita

Remember, if you need to speak to us or would just like to pop in to say hello, our office is beside the basement lounge.

ACTIVITY 6.4 Writing a Letter-Style Newsletter

1. In groups of four, write a one-page letter-style newsletter using the scenario below.

   You work at a drop-in centre for homeless youth. You have a good relationship with the business owners around the centre and part of your strategy to maintain this good relationship is to send out a monthly newsletter. Here is what you plan to include in this month's edition:

   - You have connected three teens with their families this month.
   - Thanks to a funding drive by a local high school, your "Bus Home" fund has $2300.00 in it.
   - A local dentist has donated her services for one year. She will come to the centre to perform checkups and cleanings free of charge for any youths who would like the service. She will be at the centre for three mornings a month.
   - The Salvation Army has donated exterior paint they have left over from their recent renovation. You have put together a team of youths who will begin painting this week.
   - As always, neighbours are welcome to drop in to the centre for a coffee and a tour.
   - You want to close your newsletter by thanking the community for their continued support.

   Invent all the extra necessary details to write the newsletter.

2. When your group is finished, trade your newsletter for the newsletter of another group. Edit the group's newsletter for style, grammar, punctuation, and sentence structure.

Newspaper-Style Newsletters

Newspaper-style newsletters vary in length. They are organized into sections or stories, and follow the format of newspapers. They are often sent out by schools or community centres. In the community services field you may send these to clients' parents, sister agencies, other related facilities, or agencies that send clients to you. You may also send them out to neighbours of your facility.

Here is a sample of a newspaper-style newsletter:

Happy Spring!

We know you've all been enjoying the warm spring weather, and you'll be happy to hear that we will be putting our patio furniture out this weekend. If anyone has some free time on Saturday morning at 10:00, you can meet us on the patio and help out. We'll serve a snack afterward.

Lost Shovel

Does anyone know what happened to the brown snow shovel that used to be in the front lobby? It has been missing since the last storm in March. If you used it to clear your walkway, please put it back in the front lobby. Thanks!

New Club

Mike from apartment D is starting a checkers club. Game night will be every Tuesday from 7:00 to 8:30 in the basement lounge. Mike would like you to visit him at his apartment if you want to join the club. The games start this Tuesday.

Garden Club

Our garden club will be starting up soon. Please put your name on the list in the lobby if you would like to join the club. No experience necessary — just come and play in the dirt! Once again, we will be renting a van to take a trip to Rose's Nursery for new flowers. We could use your help choosing from all the wonderful plants at Rose's.

Source: Jupiter unlimited

Don't forget to put your clocks ahead one hour next Saturday!

Your management team,
Hermes, Kathy, Frank, and Nita

Remember, if you need to speak to us or would just like to pop in to say hello, our office is beside the basement lounge.

Research Citations

For both brochures and newsletters, you may have occasion to use material from other sources. Remember that you must always give credit when you borrow from another source. This includes the use of statistics, images, and ideas. When in doubt, give credit.

Chapter Summary

- Brochures and newsletters are effective communication tools used to share information with your clients and the community.

- Brochures and newsletters must be well organized and well written to allow the reader to understand and respond to your message.

- To design effective brochures and newsletters, you must make appropriate use of white space, lists, images, colour, and typeface.

Chapter Exercises

EXERCISE 6.1 Designing Brochures

Design brochures for the following scenarios. For each, add any necessary information to create your brochure.

1. You work at a drop-in centre for homeless adults. You are planning an afternoon foot care session. Each client will consult with a foot care specialist and receive tips for keeping their feet dry and clean when possible. Participants will be given a bottle of foot powder and four pairs of socks.

2. You work for an adult female corrections agency. You are planning a two-day parenting workshop for mothers who are ready for independent living after living from six months to one year at halfway houses. You will be covering such topics as re-establishing a bond, parenting without violence, and setting up routines. You will also teach the women how to shop on a budget and cook nutritious meals. You will provide three meals each day, and the mothers are encouraged to bring their children with them.

3. You want to start a support group for men and women recovering from narcotics addictions. The support group will be run like a social club, and you will meet once a week in a community centre room. The group is not meant to replace narcotics addictions counselling; it's simply a social club for people who would like to get together in a non-threatening environment. You will have games available and serve coffee and desserts.

4. You are a housing support officer. You would like to alert residents in one of the buildings you supervise about the proper installation and use of fire alarms. There have been a few minor cooking-fire incidents in the building, and you have discovered that not all

residents keep their smoke alarms turned on. You plan to have firefighters attend the meeting. After the meeting, interested residents can have firefighters inspect their apartments for fire hazards. You will serve refreshments at the meeting.

EXERCISE 6.2 Writing Newsletters

In groups of four, put together newsletters for the following situations. You will have to develop the story ideas for each situation.

1. You work in an open custody youth home. Your home has been getting the youth to write a monthly newsletter for many years. It is time to produce the next newsletter. Write a newspaper-style newsletter that includes the following items:

   • The front porch has been repaired, and you are looking for volunteers to repaint it.
   • Parent appreciation night is approaching. Residents need to hold a meeting to discuss the night's agenda.
   • Congratulations to residents Shani R. and Jennifer G. for completing their last grade 10 high school credits.
   • This month's outing will be to the theatre to see a movie.
   • Add one more item of your choice.

2. You work in a group home for troubled youth. Four times a year you send out a newsletter to the parents, guardians, and caregivers to advise them of what is happening at the home. Write a letter-style newsletter that includes the following items:

   • All beds in the home are full, and the current mix of youth is a good one: friendships are developing.
   • You have been running a popular series of cooking lessons, and the youth are starting to take interest in cooking their own meals.
   • The federal government has renewed funding for the next fiscal year. Money has been budgeted for a new sun porch to replace the old deck.
   • Add one more item of your choice.

EXERCISE 6.3 Using Capitalization

Circle the correct choice from each italicized grouping.

1. The *college/College* will be closed over the *Winter/winter* break.

2. I discovered a shortcut while driving *south/South* on Highway *101/highway 101*.

3. I think the new addictions shelter is on *edina street/Edina Street/Edina street*.

4. The *City Planners/city planners* have granted permission for our renovation.

5. I plan to go to the *Fall Festival of Colours/fall Festival of Colours* show at the arena this Sunday.

6. The *pope/Pope* will be touring Italy next month.

7. Zamir starts his new job next *monday/Monday*.

8. I saw a great film called *Searching for Simplicity/Searching For Simplicity/searching for simplicity* last night.

9. I am dreading my dentist appointment with *doctor Aliu/Doctor Aliu* tomorrow.

10. Do you like my philosophy *professor/Professor*?

TEMPLATE 6.1   Sample Brochure

Side A

Why Meditate?	Course Details	After completing this course, you will be able to
This one-month course will teach you how to use visualization and meditation to achieve peace of mind and a positive outlook. By practising therapeutic meditation, you will learn to focus on your inner strength during times of stress.	This free course is open to adults who are currently attending or have recently completed alcohol or drug addiction programs. Each two-hour class will begin with a 30-minute lecture on visualization or meditation techniques. The lecture will be accompanied by short video clips. You will then work on individual visualization and meditation for 60 minutes. The night will end with a 15-minute group discussion. Refreshments will be served during the discussion period. Please wear loose, comfortable clothing.	• Recognize your stress triggers. • Use visualization to reach a state of inner calm. • Focus your mental discipline to overcome negative temptations. • Practise peaceful meditation to relieve stress.

Side B

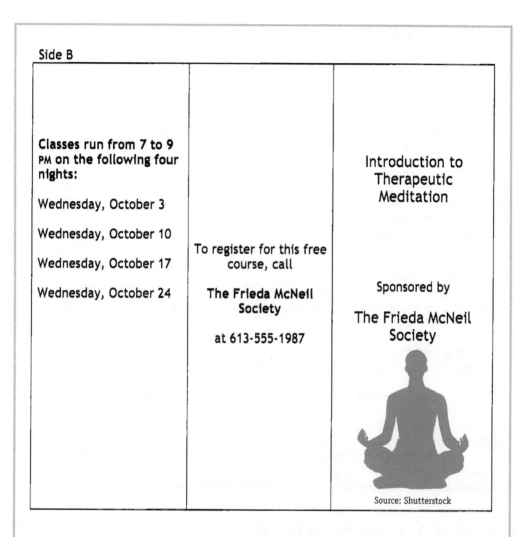

Classes run from 7 to 9 PM on the following four nights:

Wednesday, October 3

Wednesday, October 10

Wednesday, October 17

Wednesday, October 24

To register for this free course, call

The Frieda McNeil Society

at 613-555-1987

Introduction to Therapeutic Meditation

Sponsored by

The Frieda McNeil Society

Source: Shutterstock

TEMPLATE 6.2 Sample Newsletter

The Freida McNeil Society - Fall Newsletter

To Our Friends and Neighbours
Thank you for your continued support. Since our centre has moved into this neighbourhood, we've received countless good wishes and developed strong relationships with the community. We hope you continue to drop in and visit—our coffee pot is always on!

Music in the Park
Please join us at McKellar Park on Friday, September 19 for a night of blues music! A variety of Wellsley's finest musicians have donated their time to give you an evening you won't forget.

This is a pay-what-you-can event, and the money we raise will help stock our pantry and buy necessities for our clients.

Bring a picnic and something to sit on. The show starts at 7:00 PM.

Donations Gratefully Accepted!
We are still accepting donations of gently used toys and novels to restock our playroom and library. You can drop your donation at the centre at 505 Byington Street any day between 8:00 AM and 4:00 PM. Thank you.

Provincial Government Follows Through
We are pleased to tell you that the provincial government has agreed to fund our much needed renovation. In the spring, we will be getting a new roof, a new entrance that is wheelchair accessible, and a new back deck.

Gratefully yours,

The Staff

Presentation Skills

After reading this chapter and completing the exercises in it, you will be able to

1. Prepare an organized presentation targeted to your specific audience.

2. Use appropriate visuals to enhance your presentation.

3. Use body language and an excellent vocal quality to deliver a strong message.

4. Deliver professional presentations for a variety of circumstances.

Introduction

Think about the possible circumstances in which you will use your oral communication skills in the community services field. You will be working with clients, peers, health care professionals, family members, and justice professionals, to name a few. You will have occasion to speak on behalf of your clients at hearings and conferences. You may need to present funding proposals to funding boards. You may be responsible for staff training. You may have to present at an inquiry. And, of course, prior to joining the community services workplace, you will prepare and deliver presentations while you are a student.

Most people do not enjoy giving presentations. In fact, public speaking is often ranked as a person's number-one fear—a fear ranked higher than a fear of death or a fear of heights! Most people will admit that their presentation anxiety arises primarily from their dread that audience members will judge them negatively. The best defence against this fear is to be well prepared for every presentation you give, regardless of whether you are presenting an item at a staff meeting, giving testimony for a client in court, or speaking to a large audience at a conference. A well-prepared speaker will always earn the audience's respect.

> **Note:** Everyone understands presentation anxiety. If your audience can see that even though you are nervous, you are doing your best to give them the best presentation possible, they will applaud you and give you the respect you deserve.

This chapter will prepare you to be a confident speaker by giving you the tools to plan and execute a solid presentation.

SENTENCE POLISHING SKILL: MODIFIERS

A modifier is a word that gives more information about another word.

Example

I wrote a <u>great</u> report.

The word "great" is used as a modifier to describe the report.

Modifiers help make your writing precise and descriptive. You can make two mistakes, however, when using modifiers:

1. A misplaced modifier modifies an unintended word or phrase because it appears in the wrong spot in the sentence. This results in confusion and sometimes amusement. Modifiers should be placed as close as possible to the words they modify.

   Example

   <u>Barking in excitement</u>, we watched the puppies play together.

   This sentence reads as though the human beings were barking in excitement. The phrase "barking in excitement" is the modifier, and it is meant to give more information about the puppies. The phrase must be moved:

   We watched the puppies <u>barking in excitement</u> as they played with each other.

2. A dangling modifier does not modify the correct word or phrase because the subject it should be modifying does not appear in the sentence.

   Example

   <u>Before going to a movie</u>, the dog has to be walked.

Does the dog plan to go to a movie? If not the dog, then who is going to the film?

The example sentence can be fixed this way:

Before <u>we</u> go to the movie, the dog has to be walked.

or

Before going to a movie, <u>we</u> must walk the dog.

At the end of this chapter, you will find an exercise to help you practise using modifiers properly.

ACTIVITY 7.1   Giving Great Presentations

1. In small groups, take turns discussing great presenters you have heard. Think about your high-school teachers' classes, weddings, or funerals you have attended, training sessions in which you have participated, awards shows you have watched, and any other events where you have witnessed someone public speaking.

   What are some of the qualities that impress you in a speaker?

   _____

   _____

   _____

2. As a class, generate a list of the qualities of good presenters. Write down the list, and consider it whenever you are preparing to give a presentation.

   _____

   _____

   _____

   _____

Presentation Planning

As you know, a great deal of planning goes into a good presentation. This section discusses the practical things you need to consider as you design your presentation.

Awareness of Your Audience

One key to delivering a successful presentation is knowing your audience. Your first step when planning a presentation is to consider the following questions:

Audience's Size

- How many people do I expect in the audience?
- Will I need to use a microphone?

- Is the audience small enough for conducting activities and discussion groups?
- How many handouts will I need?
- How comfortable am I speaking to a group this size?

Audience's Knowledge Level

- Are the audience members experts in the field?
- Are the audience members new to the information I will discuss?

Audience's Attitude

- Is the audience sympathetic to my message?
- Will the audience need to be persuaded into considering my message?
- Is the audience hostile to my message?

Presentation Room Particulars

Another basic issue that you should assess before presenting is where you will be presenting and what sorts of equipment you will have available to you. Knowing this information will help you shape your end product.

Room Size and Set-up

- Can I move around in the room, or will I be behind a podium?
- Is the furniture moveable if I want to rearrange it to facilitate small-group discussions?

Equipment

- Will I have access to the equipment I need?
- Do I know how to use the equipment, or will I need technical help?

Length and Timing

Before finalizing your material, you need to know how much time you have to present it. As well, there may be other timing issues that will affect how you plan your content and organize your delivery.

Agreed-Upon Speaking Time

- How long will I be expected to speak?
- Is there some leeway if I go over the allotted time?
- What happens if I wrap up early?

Timing

- Am I one of many speakers?
- Where am I on the agenda?

Dynamic Presentations

Now that you know the particulars of your audience, room, and presentation length, you can plan the content, style, and flow of your presentation. Following are some organizational elements that you must develop to deliver a dynamic presentation.

Introduction

This critical portion of your presentation sets the tone for the rest of your talk. In your introduction, you must

1. Introduce yourself and your topic.
2. Interest your audience in what they are about to hear.
3. Reassure the audience that you will not waste their time.

Introduce Yourself and Your Topic

You may find this to be the hardest part of your presentation, because it will be the point at which you are most nervous. Take your time starting. Take a deep breath; take a sip of water if you need it. Then, begin in as strong and confident a tone of voice as you can muster. You will find more information about tone of voice and confident body language later in this chapter.

Get Your Audience's Interest

You can use several strategies to pique your audience's interest in your presentation. You can use

- A thought-provoking question

   Example

   Have you ever wondered what our city's homeless people do to survive the winter weather?

- Statistics

   Example

   Recent studies have shown that up to 50 percent of Canadian adults with mental health issues also have prescription-drug dependencies.

- A short anecdote

   Example

   I recently interviewed a young woman who has been living in a shelter for women fleeing from abuse . . .

Reassure Your Audience You Will Not Waste Their Time

You should show your audience you are prepared. You can

- Provide an agenda for your presentation.

   If you are giving a brief, informal talk (e.g., at a staff meeting), your agenda can simply be you saying, "There are three points I'd like to make about my proposal for a new lounge for our clients." If you are giving a longer presentation, you may want to hand out agendas or display one on a computer presentation slide.

- Tell your audience you welcome their input.

   You can do this by stating whether you will take questions during the presentation or at its end. You can give the audience your contact information in case they want to discuss your presentation with you.

- Tell your audience how listening to your talk will benefit them.

  You can provide a bulleted list of what your audience will learn/understand/be able to do after attending your talk.

- Anticipate potential questions.

  If you know your audience, you can anticipate the types of questions they may have about your presentation. For example, if you are presenting to an audience you know will be hard to persuade, you may want to have statistics or expert testimony ready to back up your material.

ACTIVITY 7.2   Introducing Yourself

In groups of four, take turns introducing yourselves as you would at the beginning of a presentation. Pick a topic, and introduce yourself and the topic in a firm, clear, confident manner. (Try choosing a topic you might have to speak about in your studies.)

Visual Aids

Visual aids provide information to your audience and can greatly enhance your presentation. They capture your audience's attention and make your talk more interesting. To be effective, your visual aids must be professional, relevant, and informative. They must also be used at the right time in your presentation.

   Possible visual aids are

- Computer presentation slides
- Handouts
- Objects
- Audiovisual media

Here are some tips for using visual aids in your presentations:

Computer Presentation Slide Show

Computer presentation slide shows are a popular way to add depth and interest to your presentation. They allow you to add photographs, graphics, and graphs. For a sample of a computer presentation slide show, please see Template 7.1 on page 123. Here are some tips for preparing computer presentation slide shows:

1. Do not put too much text on one slide.

   Computer slides add to your presentation, but they do not replace you. Ideally, each slide will contain a point or points, in bulleted text rather than complete sentences, on which you plan to expand. If you put complete blocks of text on your slides and then ask your audience to read the text, or if you read the text to them, your audience will wonder why you did not just mail the presentation to them.

2. Use an appropriate type size.

   Your slides will be useless if your audience cannot read them. Recommended type size depends on the size of the room where you are presenting; however, generally, your typeface should be at least 18 points.

3. Be careful about which special effects you use.

   Keep your slides simple. You can make your slides visually appealing without bombarding your audience with colour, movement, and graphics.

4. Limit the number of slides you use.

   The value of computer presentation slide shows greatly diminishes if you overwhelm your audience with too many slides. Aim to use no more than 10 slides in any presentation.

5. Edit your slides carefully.

   Proofread your slides several times to ensure they are free of errors. Your audience will notice spelling, grammar, and sentence structure mistakes.

Handouts

Handouts are informative supplements to any presentation. They can take the form of additional readings, summary or fact sheets, focus question lists, additional statistics or facts, or lists of resources.

Here are some tips for using handouts:

1. Keep your handouts brief. Aim for a one- or two-page handout.
2. Be ethical about your research. On the handout, provide all reference information for any material you borrowed from another source.
3. As with computer presentation slide shows, carefully edit your handout for spelling, grammar, and sentence structure errors.
4. Pay careful consideration to the layout and appearance of your handout. It should be visually appealing and easy to read.

Objects

Objects are not often used during presentations, but they are sometimes suitable. If you were presenting a proposal for the use of a new kind of panic pager, for example, you might hand around some at your presentation. If you planned a basic cooking workshop for teens, you might present samples of the food you are showing them how to make. Here are some tips for using objects in your presentations:

1. Not all participants in the room have to have their own samples of the object, but if you are speaking to a large audience, have three or four of the objects to pass around.
2. Carefully plan when you will hand out the object. When the object is being shared, your audience will be distracted by passing the object, looking at the object, and being curious about the object. This can be advantageous if you

sense you audience needs a break from your talk; however, if you are about to launch into an important part of your speech, you may need everyone's attention focused on you.

Audiovisual Media

Audiovisual clips can be very effective in presentations. As with computer presentation slide shows, they allow your audience to focus on something other than you. This will add variety to your presentation and help keep your audience engaged. We are fortunate to have access to so much electronic information; it is easy to find video and audio clips to enhance your presentation. Here are some tips for using audiovisual clips:

1. Choose your video and audio clips carefully. Use high-quality clips that are clear and large enough to be seen by all your audience members. In addition, make sure the sound quality of your clips is clear enough to be heard throughout your presentation room.
2. Make sure you know the source of the work and credit that source properly.
3. Know your material well. Make sure you are familiar with the words and actions on the clip you are using.
4. Use the clips to enhance your talk, not to replace it. Try not to succumb to the temptation to speak for 2 minutes of a 15-minute presentation and then show a 13-minute video clip. Your audience is at your presentation to hear you.

Cue Cards

Even for short presentations, using cue cards or notes can help you follow your presentation plan. Ideally, you will know your material well enough to rarely glance at the cards; however, if you are thrown off track by a question or lose your place due to nervousness, having cards as a backup allows you to quickly recover, which helps you keep up your confidence level. Here are some tips for using cue cards:

1. Number your cards. If you nervously shuffle the cards or drop them, you can easily reorganize the cards if they are numbered.
2. Use only point-form notes on your cards. This format will prevent you from reading directly from your notes. Presenters who constantly read from their presentation materials are uninteresting to watch and often speak in monotone voices. Neither of these qualities compels their audiences to pay attention to them. Furthermore, audiences attend presentations to learn about topics in a dynamic environment. Don't annoy them by reading a report or article that they could have read for themselves in the comfort of their own homes!
3. Write on only one side of the card. Although this is wasteful, two-sided cards can be confusing, especially when you are nervous.

Conclusion

Your presentation should end as strongly as it started. Your conclusion refocuses your audience and must do the following things:

- Let the audience know you covered everything you promised to cover, and remind the audience of what they learned.
- Thank the audience.

Remind the Audience of What They Learned

Give a 30- to 60-second summary of your talk's important points by briefly restating your agenda items, and point out how you covered them. This may seem repetitive, but this summary is a valuable element of your presentation. Imagine, for example, that you have just given a presentation for marks, or that you, as a paid speaker, have given a workshop to clients who are then going to be interviewed about what they learned. These audiences should be reminded that you gave a solid, informative presentation.

Thank Your Audience

There are many ways to thank your audience, depending on the nature of your presentation. You can thank your audience for their time and attention. You can thank them for asking great questions. You can thank them for participating in any audience activities you may have had them do. You can thank them for being willing to consider your proposal or point of view. Another popular concluding strategy is to leave the audience with a thought-provoking question or quote.

Tips for Giving Group Presentations

Group presentations can be very effective. To provide your audience with a professional group effort, you should consider the following guidelines:

1. Introduce all group members at the beginning of your talk. You might also want to briefly explain what the members will be speaking about as you introduce them.
2. Try to have each group member contribute equally to the talk. If a group member has a very small part or does not speak at all, your audience will question why that group member is there.
3. Each group member should remain attentive to the speaker. Side conversations among group members are distracting to the audience.
4. Group members should introduce the next speaker by saying something such as, "My colleague Eduardo will now talk to you about . . . "

Planning Your Message's Delivery

Presentations are performances. No matter what your message or content is, when you use a presentation to deliver it, you are agreeing to put on a performance. As such, there are many elements to a good performance that you must consider when designing the content of your talk.

Presentation Anxiety

There is no getting around it. Most of us must overcome, or at least control, presentation anxiety. Anxiety can be positive if you use it as a source of energy. It can help focus you on the task ahead; however, overwhelming anxiety is not helpful, so plan to do some anxiety reduction activities before your presentation.

Here are some common physical symptoms of presentation anxiety:

- Sweating
- Dry mouth
- Rapid breathing
- Upset stomach
- Diarrhea
- Uncontrollable laughing

Here are some anxiety-releasing activities:

- **Breathing exercises**
  Deep-breathing exercises slow down your breathing and your heart rate. Even doing something as simple as closing your eyes and breathing in deeply through your nose and out through your mouth five times will slow your heart rate and help calm you down.

- **Visualization exercises**
  Visualization exercises are also beneficial. One useful visualization technique you can do leading up to the day of your presentation is to visualize yourself delivering a great presentation. Not only does this help you rehearse, it also shows you that you can be successful.

- **Exercise and nutrition**
  If you exercise regularly, try to get exercise on the day of your presentation. As well, eat a balanced meal before your presentation. A well-exercised body and a well-fed brain will perform at their peaks.

- **Rehearse**
  Of course, rehearsing so you are prepared for your presentation is the best way to be as relaxed as possible during your presentation.

- **Drink water**
  If you get a dry mouth during presentations, be sure to have water with you. Taking the odd sip of water also slows you down and gives you a short break.

ACTIVITY 7.3 **Identifying Your Physical Symptoms of Presentation Anxiety**

1. Make a list of the physical symptoms you experience when you have presentation anxiety.

_____

_____

2. Share these with a partner.

3. In partners, share your tips for dealing with presentation anxiety.

Vocal Quality

To give an effective presentation it is not just your words that count but the way in which you deliver them. Ideally, you will speak in a strong, modulated, and confident tone of voice that all audience members can hear. Most of us must practise to achieve an excellent vocal quality. Here are some aspects of vocal quality to consider as you plan your presentations:

- **Volume**
  There is an art to projecting your voice without yelling. If you know you tend to speak quietly, you need to practise your speaking voice. You can do this several ways. You can hire a vocal coach. You can practise speaking in a large room, taping your voice with a tape recorder at the back of the room. After recording your voice, you can review the tape and keep practising until your voice is loud enough. You can also have a friend sit at the back of the room and give you feedback.

- **Tone**
  Everyone has been subjected to presentations by monotone speakers! If you suspect you suffer from this bad vocal habit, you should practise modulating your tone of voice. One way to add variety to your tone is to add some audience questions to your talk, which will naturally cause your tone of voice to modulate. Before presentations, you can read aloud from children's stories or adventure novels to practise a variety of intonations. You can also hire a vocal coach.

- **"Ums" and "ehs"**
  These annoying words at the end of our sentences tend to slip out when we are nervous and feel that we must fill silence. It is challenging to rid your speech of them, but it is possible. Try eliminating them in your everyday speech at work and in social situations. At first, you will consciously have to stop yourself from "umming" and "ehhing," but with practice, you will become a more polished speaker.

ACTIVITY 7.4 # Making Impromptu Presentations

Take turns with your classmates going to the front of the class and giving a one-minute talk on a topic of your choice. Speak loudly and clearly. Consciously try not to end or begin any sentences with "ums" or "ehs."

Body Language

If you are nervous during presentations, be aware of the body language you may exhibit as a result. Do you play with objects (e.g., pens, papers, paper clips) if they are in your hands? Do you fidget with your earrings or glasses? Do you play with your hair?

If your presentation content is well-prepared, you will be able to concentrate on using confident body language during your talk. Confident body language includes standing up straight, moving around in a natural manner, and smiling and using hand gestures when appropriate.

Eye Contact

In Western culture, direct eye contact is regarded as a necessary communication skill. Even though you may be nervous during your presentation, you must make the effort to maintain eye contact with your audience. You will benefit from the positive looks of reinforcement that you get in return from the positive audience members at whom you look. One common tip for maintaining eye contact is to pick a focus spot at which to look on the back wall of the presentation room. This can work if you are speaking in a large, dimly-lit lecture hall; however, in a smaller venue (e.g., a classroom, boardroom, or courtroom), if you conduct your presentation while staring at a spot on the back wall, it will be obvious that you are doing just that—staring at the wall! A better technique is to mentally divide the room into four quadrants. Look at one or two people from each quadrant, one quadrant at a time. Do this randomly; do not try to rotate equally from quadrant to quadrant because your audience, again, will notice. Within the first few minutes of your presentation, you will pick out the friendly, positive audience members. They will be looking at you and smiling or nodding. Once you have located them, return your gaze to them throughout your talk. Their support will help relax you.

Chapter Summary

- Strong presentations require thoughtful planning.

- Presentations must begin and end strongly.

- Effective presentations require appropriate body language, eye contact, and excellent vocal quality.

- Correctly used visual aids can enhance your presentation.

- You can manage your presentation anxiety by acknowledging it and decreasing it through relaxation exercises and practice.

Chapter Exercises

EXERCISE 7.1 Making Impromptu Presentations

1. Take 10 minutes to organize a talk about your career plans for the next five years. In groups of four, take turns giving a three-minute career plan talk.

2. Take turns with your classmates going to the front of the class. When it is your turn, introduce yourself and ask the audience a question about food that is linked to your favourite food. Then give a description of your favourite food.

3. In groups of four, prepare a four-minute group presentation about a sport or pastime. Give your presentation to the class.

EXERCISE 7.2 Preparing Presentations

1. Prepare a three-minute talk about someone who has inspired you. Bring in an object that reminds you of that person. Be prepared to explain the object and pass it around at an appropriate point in your talk.

2. Choose a controversial article from a magazine or journal in your field (you can use online sources). Give a five-minute talk during which you summarize the article and give your opinion about the topic.

3. Prepare a 10-slide computer slide show on a topic of your choice. Be sure to follow the chapter guidelines for producing effective computer slides. Present your slide show to your classmates for critique.

EXERCISE 7.3 Using Modifiers

Rewrite the following sentences so that the modifiers are used correctly. You may have to move around the existing modifiers or add subjects.

1. With only $3.00 left, the sweater was too expensive.

2. Susan accepted the award for bravery without drama.

3. While using the Internet, my cat escaped through an open window.

4. Accused of hiding the knife, the youth worker told her she would lose her achievement points for the day.

5. After locking the door, the smoke alarm started beeping.

6. I love rainy days when the children play in the puddles wearing shiny yellow raincoats.

7. After claiming to be bored, the dancing seal was a great hit.

8. You can use a saw to cut the large tree with sharp teeth on it.

9. Try not to move when the dentist gives you a needle with trembling hands.

10. Jack told his girlfriend that he would like to teach her to bowl keenly.

TEMPLATE 7.1 **Presentation Templates** (Sample Computer Presentation Slide Show)

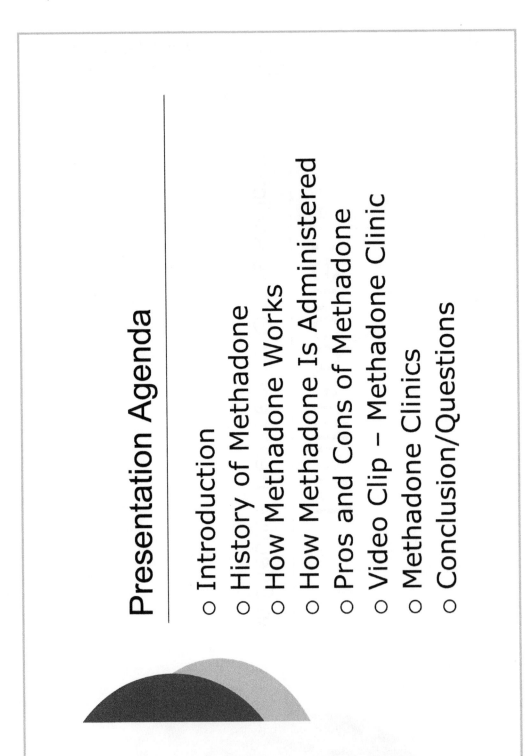

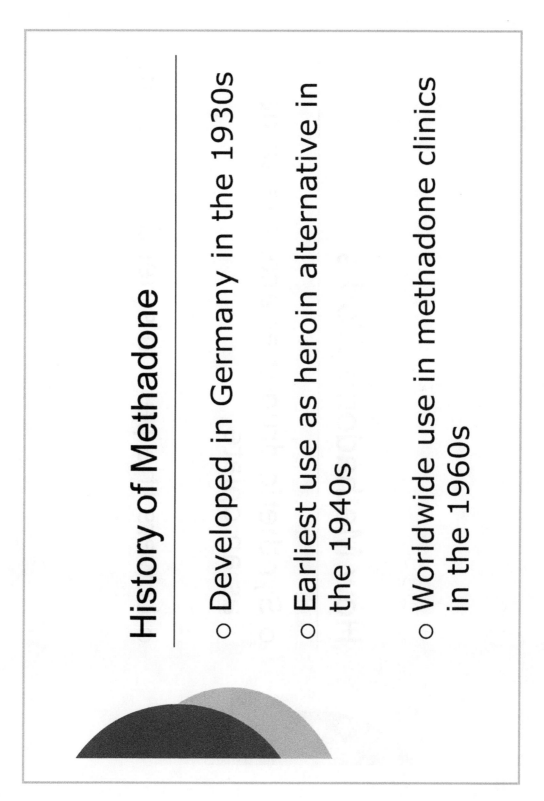

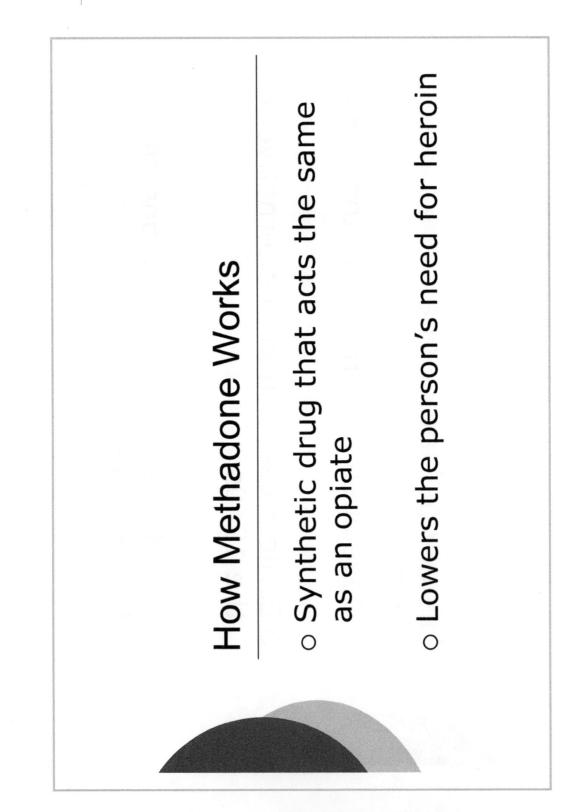

How Methadone Works

○ Synthetic drug that acts the same as an opiate

○ Lowers the person's need for heroin

How Methadone Is Administered

o Usually given in pill form

o Dose is gradually increased until recommended level is achieved

o Patients visit a clinic or pharmacy daily for dose

o Methadone is used for as long as necessary

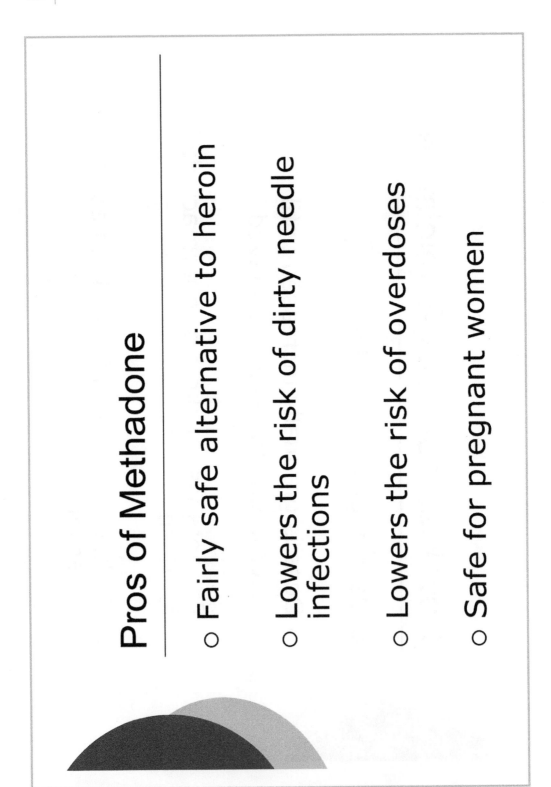

Pros of Methadone

○ Fairly safe alternative to heroin

○ Lowers the risk of dirty needle infections

○ Lowers the risk of overdoses

○ Safe for pregnant women

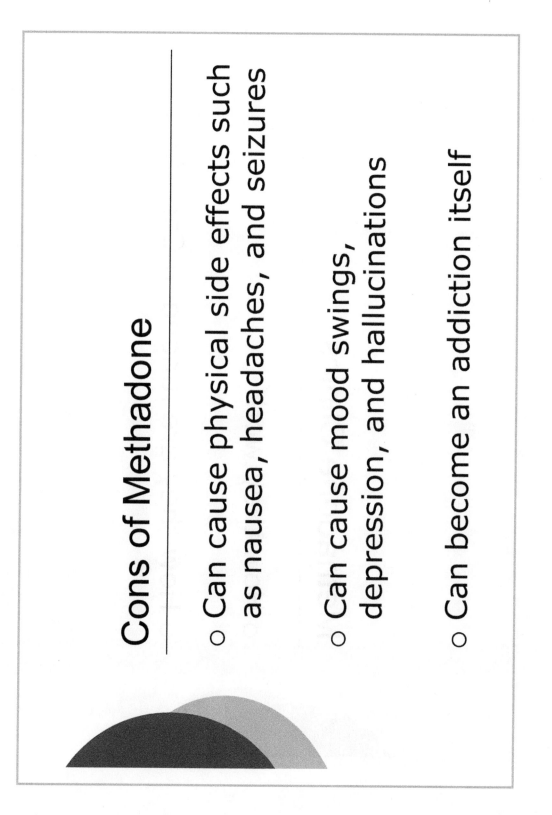

Cons of Methadone

o Can cause physical side effects such as nausea, headaches, and seizures

o Can cause mood swings, depression, and hallucinations

o Can become an addiction itself

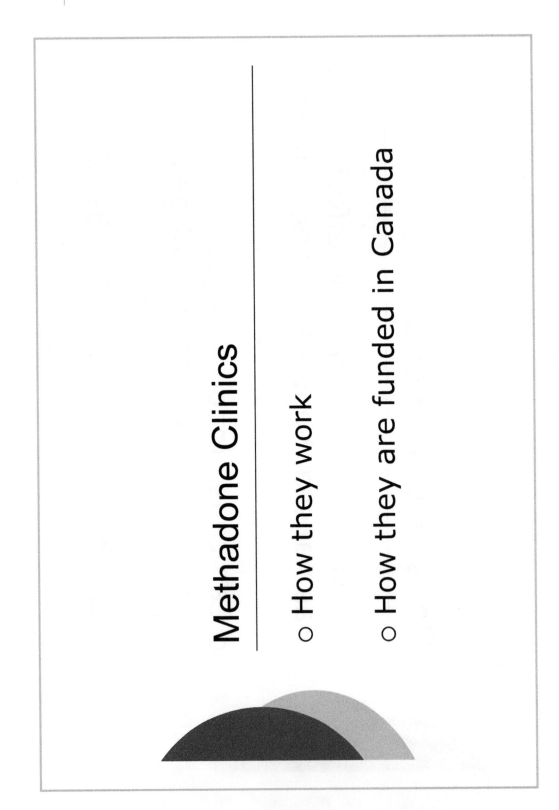

Methadone Clinics

○ How they work

○ How they are funded in Canada

References

www.drughealth/methadone.org

www.Canadianhealthwatch/
methadonealert.gov

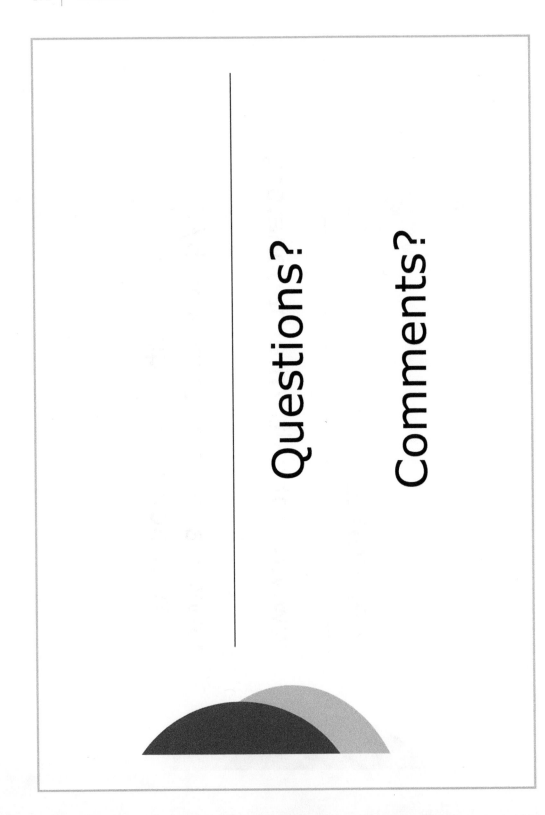

Job Search Skills

LEARNING OBJECTIVES

After reading this chapter and completing the exercises in it, you will be able to

1. Organize a professional resumé that effectively highlights your professional, educational, and personal accomplishments.

2. Write a persuasive cover letter that gets you invited to an interview.

3. Conduct confident interviews in which you articulately explain your unique skills and suitability.

Introduction

The job search process is exciting and stressful. You know you are ready to get into the community services profession and make a difference. You have researched the job market, are completing the relevant education, and may even have some experience in the field. Now you have to convince employers that you are the perfect candidate for the job. The best way to do this is to prepare a professional resumé, write a convincing cover letter, and practise a lot before your interviews.

SENTENCE POLISHING SKILL: BULLETED LISTS

Often the information you want to put in a letter, memo, report, or resumé is best organized by using bulleted lists. Bulleted lists add visual appeal, clarity, and structure to your writing. They also break up a paragraph of text by introducing white space.

Here are some examples of bulleted lists:

We are starting three new life-skills programs:

- Self-esteem for teens
- Nutrition for teens
- Time management for teens

Please improve your report writing skills by focusing on these areas:

- Using the active voice
- Attending to detail
- Maintaining a professional tone
- Editing for grammatical errors

The inmates made four demands:

- Bigger meal portions
- Extended yard time
- Reinstatement of a smoking lounge
- Increased telephone time

Here are some guidelines for using bulleted lists:

- Use parallel structure (for more information, see "Sentence Polishing Skill: Parallelism" on page 2).
- Use a symbol or bullet (such as the small black circle used here) to indicate each item on your list.
- Keep each item in the list brief when possible.
- Try to use only one bulleted list in a memo or a letter.
- Try not to put more than one bulleted list in any given section of a report or other long document.

At the end of this chapter, you will find an exercise to help you practise using bulleted lists.

Resumés

A resumé is a summary of your professional, educational, and personal accomplishments. Resumés are no longer just a chronological listing of the jobs, education, and other experience you have had. Instead, a strong resumé highlights those experiences that best showcase how your abilities and experiences make you the perfect candidate for the job you want. Hundreds of resumé templates are now available through word processing packages. You can choose one of them to prepare a well-organized and easy-to-write resumé. All you need to do is open your template of choice and insert your personal information.

Choose a business-style resumé template, and keep your layout simple by following these guidelines:

1. Use only one typeface.
2. Highlight headings and subheadings using boldface or italics. Do not use all capital letters. Many readers find words written in all capital letters difficult to read. Italics can also be hard to read, so use italics with caution.
3. Use white, high-quality paper.
4. Use black ink.
5. Keep your resumé to two pages.

In certain large organizations, you can submit your resumé only electronically. If this is the case, you may have no control over what your resumé looks like. Instead, you will be required to fill out a resumé form. In this situation, when you go for the job interview, bring some copies of your actual resumé to present yourself with a professional, polished, and personalized document.

Your resumé will start with these standard categories:

- Education
- Work experience

You will then include some or all of the following categories, depending on your experience and the type of job for which you are applying:

- Related experience
- Other skills and accomplishments
- Volunteer experience
- Licences and certifications
- Awards and commendations

More information about these resumé sections follows below.

Education

Because the education you are now pursuing relates directly to the field you are entering, you may choose to start your resumé with the education section. However, if you have experience in the community services field, you may want to begin your resumé with the work experience section to highlight it.

In your resumé, list your schooling from present to past and start with your current program of study.

Examples

Here is a sample of how your resumé will start if you are still in the program:

Child and Youth Worker Program	September 2008 – present
Harter College	
I have completed two semesters of a four-semester program. I have maintained an A- average.	

Here is a sample of how your resumé will start if you have completed the program:

Child and Youth Worker Diploma	2006 – 2008
Harter College	
(Dean's Honour List)	

Also include any other post-secondary education you have, even if you have only partially completed a degree or diploma because all completed courses count as valuable education.

Example

Business Administration	2004 – 2005
Pennant University	
I completed one term of a three-year degree. My completed courses include Introduction to Accounting and Small Business Management.	

Your high school education is also an important point to mention.

Example

Prince Edward Island High School Diploma	1998 – 2001
St. Julien High School	
(Honours)	

In your resumé, do not include

- Your elementary school education as it is a given that you completed elementary school
- Courses you started but did not complete

Work Experience

You can set up your work experience section in a few ways.

Chronological Style

Chronological style follows the same format that the educational section uses. List jobs from present to past, and include the organization, job title, and period of employment. You can follow this information with either a bulleted list or a brief paragraph that describes your duties and accomplishments. You could also have a bulleted list *and* a brief paragraph. For a sample of a chronological-style resumé, please see page 140.

Functional Style

For this resumé style, organize your experience into skill sets rather than by jobs. For each section, explain how and where you learned the skill. Organizing your job experience this way puts the emphasis on your skill sets, not your places of employment. At the end of your skill sets section, you can list your past employers.

When listing your skills, choose words directly from the job advertisement or job description for which you are applying. For example, if the job requires strong communication skills and proven experience working as part of a team, two of your skill sets should be communication skills and teamwork skills. To round out those skills, choose other skills you feel are important for the job. Here are some suggestions for skills needed in the community services field:

- Advocacy skills
- Communication skills
- Counselling skills
- Report writing skills
- Research skills
- Independent work skills
- Interview skills
- Leadership skills
- Teamwork skills
- Writing skills
- Negotiation skills
- Oral communication skills

Aim to highlight three to five skill sets for your resumé.

If you have had many jobs and do not want to list them all on your resumé, or if none of your work experience relates directly to the job for which you are applying, the functional-style resumé is an effective way to highlight your qualities.

For samples of functional-style resumés, please see the resumé at the end of this section on page 142 and Template 8.2 on page 153.

Action Verbs

Use action verbs to describe your work experience and accomplishments.

- Authored
- Controlled
- Counselled
- Designed
- Facilitated
- Implemented
- Initiated
- Innovated
- Invented
- Led
- Liaised
- Mediated
- Mentored
- Motivated
- Negotiated
- Organized
- Planned
- Reviewed
- Supported
- Trained

Bullets

Use bullets to organize your information. If suitable, you can use a combination of bullets and a short descriptive paragraph.

Here is a sample of the use of bullets and a descriptive paragraph:

Molly's Bargain Barn   2003 – 2005

Manager

- Hired, trained, and supervised part-time staff
- Planned and assigned weekly staff schedule
- Prepared weekly bank statements
- Designed window displays

I started at Molly's Bargain Barn as a part-time cashier and was promoted to head cashier after four months. After six months as head cashier, I was promoted to manager. As manager, I implemented a new inventory system that saved the company, and I was given a commendation. I left to start the Social Service Worker Program.

When choosing how to describe your duties or accomplishments at a job, consider what unique skills you learned and what special contributions you made while on staff. Every employer knows the basic duties of a retail manager. What can you tell employers about what *you* brought to the job?

Choose from the following sections to complete your resumé. You may have information to put in each section, or you may choose to add only one or two of these sections.

Related Experience/Volunteer Experience

Choose one of these titles to list all the volunteer work you have done, including the hours you completed as part of your high school graduation requirements, if applicable.

Other Skills and Accomplishments

Use this section to describe such things as membership on sports teams, participation in charity events, and enrolment in specialty training.

Licences and Certifications

You never know when a special licence or certification may be the clincher that gets you a job.

Use this section to list such things as a specialized driver's licence, a bronze medallion in swimming, a whitewater rafting certification, a restraint techniques certification, a chainsaw licence, aerobics instructor training, or a small-engine repair certification.

Awards and Commendations

Awards and commendations show your history of striving for excellence. If you have been recognized for participation on a team or in a contest, for your role as a leader or innovator, or for outstanding performance, put that information in this section.

Objectives

There was a trend in the late 1980s to begin every resumé with an objective—a one-sentence description of your career goals or intended path. This was a useful tool if you were applying for a job in the business sector.

Example

Objective: To begin my accounting career at an established accounting firm and, through hard work and training, advance according to my potential

Starting resumés with an objective is not as popular today, because people realize that the technique is not effective for every work situation.

Consider these examples:

Objective: To be hired by your police service as a police officer
(No kidding!)

Objective: To be hired at Serenity House until I have some field experience, at which point I plan to move to a larger facility
(Who wants to train someone who wants to switch jobs as soon as possible?)

Use the objective opening if you are applying to a large organization or facility. Do not use the objective opening if you are applying to a small centre, halfway house, or other community services facility.

Hobbies

Save hobbies and other types of personal information for the interview. For example, if you sew or do woodworking, you may mention this during an interview at an assisted living residence as something you can share with residents.

References

The common practice is to write "References available upon request" on the bottom of your resumé.

Do not include names, numbers, or reference letters with your original application unless you are specifically asked to do so. Employers do not call referees until they have made the decision to hire you. There is no time to call the referees of every applicant on the chance that the applicant will do well in the interview and will be offered a job. You should bring your reference list or letters of recommendation with you to the job interview. You should also bring extra copies of your resumé with you.

ACTIVITY 8.1   Preparing a Skills Inventory

Choose three or four jobs you have had, and list them across the top of the following chart. You can choose from paid and volunteer jobs. Then, fill out the rest of the chart. Show your chart to a partner. Each of you should ask questions about each other's jobs to see if you can discover additional skills that are not already on your lists.

	Job 1	Job 2	Job 3	Job 4
What were/are your duties at this job?				
What skills have you learned at this job?				
What contributions or innovations have you made at this job?				

You now have an inventory to use when you write your resumé and cover letter.

Here is a sample of a chronological-style resumé:

Regina Peters   **57 Percy Place, Halton, Manitoba ROH S2E 204-555-1298**

Education

Child and Youth Worker Program   **September 2008 – present**
Halton College

I have completed two semesters of a four-semester program. I have maintained an A- average.

Business Administration   **2004 – 2005**
Pennant University
I completed one term of a three-year degree. My completed courses include Introduction to Accounting and Small Business Management.

Manitoba High School Diploma   **1998 – 2001**
St. Julien High School
(Honours)

Experience

Molly's Bargain Barn   **2004 – 2008**
Manager
• Hired, trained, and supervised part-time staff
• Planned and assigned weekly staff schedule

- Prepared weekly bank statements
- Designed window displays

I started at Molly's Bargain Barn as a part-time cashier and was promoted to head cashier after four months. After six months as head cashier, I was promoted to manager. As manager, I implemented a new inventory system that saved the company time and money, and I was given a commendation. I left to start the Child and Youth Worker Program.

Independent Auto Sales Receptionist   2003 – 2004

- Answered phones and made appointments for sales reps
- Typed monthly sales reports
- Designed and wrote monthly flyers for sales events

This company closed in 2004. During my 10 months at the company, I was frequently praised for my writing skills and was often asked by sales reps to write letters and reports for them. During closing procedures, the sales manager asked me to write letters to clients and other auto sales companies to advise them of the closing.

Kelly's Place   2001 – 2003
Front-Line Support Staff

- Made snacks for the youth
- Played games and did crafts with the youth
- Listened to the youth when they wanted to talk about issues

I loved this experience, but I did not have enough education to be considered for a full-time position. I was advised by the director that if I wanted to pursue a career in human services, I should be educated in the field. She also told me that once I was, she would like me to apply at Kelly's Place.

Accomplishments

- Honours Math Award, Grade 12, St. Julien HS
- MVP, Senior Girl's Floor Hockey St. Julien HS
- Honourable Mention, Stanton Poetry Contest

Accreditations

- Bronze Medallion
- First Aid
- WHIMIS

References available upon request.

Here is a sample of a functional-style resumé:

Regina Peters   57 Percy Place, Halton, Manitoba R0H S2E 204-555-1298

Education | **Child and Youth Worker Program**   **Sept 2008 – present**
Halton College

I have completed two semesters of a four-semester program. I have maintained an A- average.

Business Administration   **2004 – 2005**
Pennant University
I completed one term of a three-year degree. My completed courses include Introduction to Accounting and Small Business Management.

Manitoba High School Diploma   **1998 – 2001**
St. Julien High School
(Honours)

Experience | **Writing Skills**
I have been complimented on my writing skills at high school and all my jobs. As well, my teachers at Halton College have complimented my report-writing abilities.

- At my retail job, I wrote employee performance appraisals that my supervisor used as references for writing his appraisals.
- At my job at the auto sales outlet, I wrote many letters and reports.
- At a youth drop-in centre, I had the opportunity to write shift logs and incident reports. I was told that my reporting style was direct, informative, and professional.

Organizational Skills
- As a manager I arranged schedules, ordered inventory, and organized monthly sales. My supervisor commended me for designing the smoothest staffing procedure with which he has worked.
- At my job at an auto sales outlet, I was in charge of organizing the appointments for 23 sales staff.
- While at a youth drop-in centre, I reorganized the games room and developed a nightly chores schedule for the part-time staff. I designed the chores schedule because various part-time staff members had been complaining to me that certain staff members were not pulling their weight. I did both these activities on my own initiative and was praised by the director.

Oral Communication Skills
Throughout my work experience I have been praised for my oral communication skills.

- As a manager, I ran staff meetings, counselled employees when disagreements arose, and effectively liaised with customers who had queries or complaints.
- At my job at an auto sales outlet, I talked to many clients, and several times clients commented that my service and professional attitude was what convinced them to make an appointment with a sales rep.
- My director at a youth drop-in centre complimented me often on my communication skills. She told me I had a natural way with clients and should put my oral communication skills to use in the human services field.

Employers

Molly's Bargain Barn	2004 – 2008
Independent Auto Sales	2003 – 2004
Kelly's Place	2001 – 2003

Accomplishments

- Honours Math Award, Grade 12, St. Julien HS
- MVP, Senior Girl's Floor Hockey St. Julien HS
- Honourable Mention, Stanton Poetry Contest

Accreditations

- Bronze Medallion
- First Aid
- WHIMIS

References available upon request.

Cover Letters

Cover letters are challenging to write. They are essentially a sales pitch for you. Competition for jobs is fierce, and your cover letter is the tool that will get you to the interview stage of the hiring process. For samples of cover letters, please see the cover letters below and Template 8.1 on page 152.

Consider the following cover letter:

---

17 Freemason Dr.
Redville, MB R0H 4R4

February 28, 2009

Youthsave House
223 Markland Rd.
Redville, MB R0H 4R3

Dear Norm Bruce,

My name is Batul Singh, and I am applying for the position of full-time front-line worker as advertised in the *Redville Courrier*. I am currently a student in the Social Service Worker Program at Plains College, and I believe the courses I've completed at the college will make me a strong candidate for your team.

According to your job advertisement, you are looking for a social service worker who can work independently while running programs and handling a caseload of youth. I am taking courses in the Social Services Program that give me experience with designing programs. I am a strong writer with very good report writing skills. I have been doing volunteer work in the social services field for the past three years and have worked independently in a number of roles.

Thank you for considering my candidacy. I would be happy to meet with you to discuss my suitability as a member of your team. I can be reached at 204-555-7689.

Sincerely,

Batul Singh

This cover letter is professional: it is well written; it contains no errors; it is well organized on the page; and it gives many details about the writer. However, it is also generic, not terribly convincing, and missing specific selling points.

Here is a sample of another cover letter by Batul Singh. She is applying for the same job with the same credentials.

17 Freemason Dr.
Redville, MB R0H 4R4

February 28, 2009

Youthsave House
223 Markland Rd.
Redville, MB R0H 4R2

Dear Norm Bruce,

My name is Batul Singh, and I am applying for the position of full-time front-line worker as advertised in the *Redville Courrier*. I am an A student in my second year of the Social Service Worker Program at Plains College, and I believe the courses I've completed at the college, along with my work and experience, make me a strong candidate for your team.

According to your job advertisement, you are looking for a social service worker who can work independently while running programs and handling a caseload of youth. I am taking courses in the Social Services Program that give me experience with designing programs. The self-esteem workshop I developed in school won the Dean's Award for Innovative Thinking, and the program is now being used at Micheal's Place, where I currently volunteer. I am a strong writer with very good report writing skills. I have maintained an A+ in my communications course, and my professor has used several of my reports as examples in class.

I work at a local music studio as a singing coach for teens ages 13 to 16. I independently plan lessons and coach clients. As a volunteer at Micheal's Place, I am team lead for the weekend shift. My duties include supervising five male youths with behavioural challenges.

Thank you for considering my candidacy. I know my education and experience would make me an excellent fit with your team. I would be happy to meet with you to discuss my suitability as a member of your team. I can be reached at 204-555-7689.

Sincerely,

Batul Singh

Set up your cover letter using the standard letter format discussed in Chapter 5 on page 74. Then, write each body paragraph.

Introductory Paragraph

- Introduce yourself and get your reader's attention by showing how you are the perfect candidate for the job:

  My name is Lorna Wallace, and I am an honours graduate from the Child and Youth Worker program at Orleans College.

- Clearly state you are applying for the job:

  Please consider me for the position of Drop-in Centre Coordinator.
  or
  I am applying for the position of Drop-in Centre Coordinator.

- Explain how you came to know about the job:

    I read about this position in yesterday's newspaper.
    or
    Your Director, Randy Pines, told me about an opening at your facility.

Body Paragraph(s)

- This paragraph must engage, impress, and convince your reader. Use specific vocabulary to describe your skills, strengths, experience, accomplishments, and unique attributes. You may need more than one paragraph for this part of the letter, but keep your letter to one page.
- Use the wording from the job advertisement. If you are responding to an advertisement that asks for someone with leadership experience, write something like

    I have three years of leadership experience in school council and summer camp.

- Back up all claims with details:

    I am a proven team player. I have been voted most sportspersonlike player for my ringette team for the past three years. When engaged in group work at college, I am usually asked by my peers to take a lead role. I have been complimented on my ability to get along with all students. I thrive on being part of a successful team and look forward to improving my team skills in the community service field.

Cover letters are daunting to write. Your first draft may have too much detail and will need to be pared down. Remember, your goal is to keep any business letter to a single page.

To make writing cover letters easier for yourself, prepare and save a bank of skills sentences and paragraphs that describe the variety of skills, accomplishments, and experiences you have. Add to the list as you get more experience.

Whenever you apply for a job, refer to the list. For example, if you are applying for a job that requires leadership and initiative, examine your examples that discuss these skills.

Concluding Paragraph

- Thank the reader for considering you.
- Remind the reader that you are the best candidate for the job.
- State that you will be calling to confirm the reader's receipt of your resumé and cover letter, or tell the reader you look forward to receiving a phone call from him or her.
- Include your contact information.

ACTIVITY 8.2 Listing Your Unique Attributes

On a piece of paper, list three of your *unique* attributes. These can be character traits (e.g., level-headedness, inquisitiveness, loyalty). They can be accomplishments (e.g., a scuba licence, knowledge of sign language, leadership training). They can be skills (e.g., cooking, shorthand, small engine repair). Take turns with your classmates reading out your attributes. Cross out any characteristics someone else also shares. If you cross out all your qualities because other people share them, you need to think more about what makes you special. Do not worry; you are unique. You may just need friends or family to help you identify your special qualities.

Interviews

Job interviews are certainly one of the most challenging and nerve-racking experiences through which you can put yourself. If you are prepared and have a positive attitude going into the interview, however, you can enjoy the challenge with confidence.

Before the Interview

Start preparing for the questions your interviewers will ask. Interviews are often conducted by small panels of interviewers, and you can expect to be asked a combination of traditional-style, behavioural-style, and situational-style questions. While you cannot predict with certainty what your interviewers will ask, you can use your knowledge of the field to develop possible interview questions. You can also ask people in the field for their input on potential interview questions.

Traditional-Style Questions

Interviews typically start with a few easy-to-answer traditional-style questions:

- What have you done to prepare for a career in this field?
- What makes you think you will do well in this field?
- Can you describe one of your weaknesses and tell us what you are doing to fix it?

Behaviour-Style Questions

Employers are certainly interested in your education and work experience. They are what got you the job interview. However, employers also want to know what kind of person you are. They wonder if your personality makes you a good fit for their field and their team.

One way employers get this information is by asking behaviour-style questions. Behaviour-style questions ask you to demonstrate desired behaviours by describing an experience in which you have exhibited that behaviour trait.

Examples

Working as part of a team is crucial at this facility. Can you tell us of a time when your membership on a team resulted in a positive change or outcome?

Empathy is an important trait in this field. Can you tell us of a time when you demonstrated empathy, and what the result was?

To answer these questions thoroughly

1. Describe the situation.
2. Describe your actions.
3. Explain the result.

Here is a sample answer for the second behavioural-style question given above:

> **(situation)** One day this past September I was coming into school for a class. I noticed a person with a visual impairment standing to the side of the door of the building I was entering. The woman had a large schoolbag, a cane, and a cup of coffee and seemed to be hesitating in front of the door. I asked the woman if I could help her. She thanked me and said that she couldn't tell where the door handles were. I thought about just opening the door for her, but I remembered how helpless I felt a year ago when I broke my leg. I wanted to be independent but people kept helping me even when I did not ask for help. I know how frustrating it is to be helpless. **(action)** I described the location and style of the handles. The woman was then able to open the door. **(result)** She thanked me again and said that most people would have just opened the door. She was happy that she could now open the door on her own.

Situational-Style Questions

For situational-style questions, interviewers will give you a situation and ask how you would handle it. Again, from your answers, the interviewers are looking at your behaviour or character. They want to know you are the right *type* of person for the job.

Examples

How would you deal with a client who begged you to overlook the fact that she missed curfew by five minutes so she wouldn't lose her phone privileges?

What would you do if you caught a colleague taking medication for her own use from the client medication drawer in the office?

To answer these questions thoroughly

1. Identify the problem.
2. Describe the options available to you.
3. Explain which option you would choose and why.

Here's a sample answer for the second situational-style question given above:

> **(the problem)** Taking anything that is not yours is wrong. **(options)** If I caught a colleague stealing medications, I would have a few options. I could report the colleague to my supervisor. I could talk to the colleague and ask her to put the medication back. I could pretend I didn't see the theft. I could also confront the colleague and tell her I would give her a chance to report herself to the supervisor. I would tell her that if she didn't do that within a certain time frame, I would have no choice but to tell the supervisor myself. I would choose this last option. **(why this option)** By handling the situation this way, the colleague knows I've given her the chance to help herself. I think this is a supportive stance. I also know that it would be unethical of me to do nothing, which is just not an option for me. I would also tell my colleague that I was there to help her and support her. I would tell no one about the situation other than, if necessary, the supervisor.

Other Interview Preparation

After a company contacts you for an interview, you should research the organization, prepare your employment-related documents, ensure you know where the interview is taking place, and plan your attire.

1. **Research the organization.** Interviewers will often want to know why you want to work for them. They want to know what made you apply for the job and why you think you will be a good fit.

   To prepare your answers to these questions, look at the organization's Web site, talk to people who work at the company, and read publications about the organization. Find out about the organization's history, mandate, size, client base, and main functions. You can also search news sites to see if the organization has been in the news.

2. **Prepare a folder** to bring with you to the interview with the following contents:
   - Copies of your resumé
   - Copies of your reference list
   - A piece of paper with the name(s) of your interviewers on it
   - A list of questions you want to ask the employers

3. **Verify that you know** where the interview is and how long it will take you to get there.

4. **Make sure you have the appropriate attire.** You should plan to wear a suit to your interview unless your contact has specifically told you to dress casually. Even if the employees at the organization at which you are interviewing do not wear suits on the job, it is still customary to arrive for the interview in professional dress.

On the day of the interview, before you leave your home, go over this checklist:

- I am tidy, clean, and professional.
- I know where I am going and how long it will take to get there.
- I know how to pronounce the names of my interviewers.
- I have a folder with copies of my resumé and my references list.
- I will enjoy myself by taking some control of the interview. As I answer questions, I will also think about whether I like the people and the organization and whether I would accept a job there.

During the Interview

To keep yourself focused and confident throughout your interview, follow these guidelines:

1. Introduce yourself with confidence, and do your best to smile and project a positive attitude.
2. Relax in your chair but maintain good posture and eye contact.
3. If you need to take your time with an answer, do so.
4. If you need a question repeated, clarified, or explained, ask for that.
5. **Be prepared with questions.** At the end of an interview, it is customary for employers to ask interviewees if they have any questions. Something may come up during your interview about which you would like to ask. Here are some other suggestions: Do you

offer professional development opportunities? What are some of the organization's future goals/plans? What is your most typical client profile?

Do not ask questions about salary, bonuses, vacation, or office space.

6. Thank the interviewers before you leave.

After the Interview

A debrief after any challenge allows you to acknowledge your success at meeting the challenge. It also allows you to critique your performance for any needed improvement. Here is how you can debrief after your interview:

1. Congratulate yourself for facing the challenge.
2. Send a brief thank-you email or note to the interviewers.
3. If you do not get the job, ask the employer where you went wrong. If you made mistakes in the interview, you can improve. If you were a strong candidate but not quite as qualified as the successful candidate, asking about your performance will give you the opportunity to ask the employer to keep you in mind for a future opening.

Chapter Summary

■ A strong resumé is written in a style that best highlights your experience and credentials.

■ Cover letters should convince employers to grant you an interview. The letters must be persuasive, unique, and professional.

■ Cover letters must be tailored to fit the circumstances of each job opportunity.

■ In order to be prepared for an interview, candidates must invest time in researching the employer and practising possible interview questions.

■ A strong interview candidate is well dressed, confident, and ready to face the interview challenge.

Chapter Exercises

EXERCISE 8.1 **Using Bullets in a Resumé**

Using the same jobs you charted for "Preparing a Skills Inventory" on page 139–140, write resumé entries. For each job, use a bulleted list to describe your duties or accomplishments.

EXERCISE 8.2 ## Writing a Cover Letter

1. Use the following job advertisement to write a cover letter. Use your own education, employment experience, and other accomplishments. Try to incorporate some of the vocabulary from the advertisement into your cover letter.

Youth Drop-in Centre

We are looking for part-time front-line staff members to work evenings and weekends at our drop-in shelter for homeless youth.

Front-line weekend and evening staff members tidy the lounge, kitchen, and office; answer phones; and answer questions the youth might have about available services. The staff member is also expected to spend time in the lounge keeping the youth company and playing games with them. Some kitchen duties are also included. Interested candidates must be independent workers with strong communication skills. Education or work experience in the field is an asset. Candidates must have a criminal records check.

Interested candidates are asked to apply to

Marianne Hodges at
673 Centre St.
Wellville, QC J1V G6H

---

2. Get into groups of four. Trade letters with another group. Critique the letters using these criteria:
   a. Is the letter well organized?
   b. Is the letter persuasive?
   c. Does the letter describe unique skills, accomplishments, or credentials?
   d. Is the letter error free?
   e. Would you ask this person for an interview?

EXERCISE 8.3 ## Practising Interviews

In groups of four, draft four situational-style and four behavioural-style questions. Give each group member a situational-style and a behavioural-style question to answer. Take 20 minutes to individually write answers for the two questions assigned to you. Then, get together with your group members, and take turns sharing your answers. Discuss the strengths and weaknesses of each answer.

TEMPLATE 8.1 Sample Cover Letter

34 Winding Way
Ottawa, ON K2P P2Y

October 2, 2009

Mrs. Penny Scott
East End Seniors' Social Centre
436 Willow Ave.
Ottawa, ON K8Y 1R8

Dear Mrs. Scott,

Please consider me for the part-time social facilitator position at the East End Seniors' Social Centre. Judy Nicholson, my supervisor at Sunset Seniors Home, told me about this job. I believe my education in social services and my experience working with seniors will make me a valuable addition to your team.

I have just started my second year at the Social Services program at Valleyview College where I maintain an A average. Some of the relevant courses I have completed are group dynamics, communications, and senior care management. In my communications class, we learned effective conversation techniques to use with clients who have Alzheimer's disease, which is one of the skills you listed as an asset in your job posting. While attending school, I have been volunteering at Sunset Seniors Home. My main responsibility at the home is to be a companion to the residents. In this capacity, I chat with residents, accompany them on walks, work with them in the crafts room, and play board games with them. Your job posting emphasized your search for candidates who are sociable and show initiative. I believe I exhibit these traits. For example, in May, I started an Agatha Christie book club at the home, and I am proud to say it is very popular.

Thank you for considering my suitability for the position. I look forward to hearing from you and can be reached at 613-555-7834 or at Bennett.t@homenet.com.

Regards,

Tanya Bennett

Tanya Bennett

TEMPLATE 8.2 Sample Skills-Based Resumé

34 Winding Way
Ottawa, ON
K2P P2Y

Phone 613-555-7834
Email
bennett.t@homenet.com

Tanya Bennett

Education

2008 – present Social Services Program, **Valleyview College**
- Dean's Honour Roll

2005 – 2008 **Allcott High School**
- Principal's Honour Roll

Functional summary

Communication Skills

My paid and volunteer experiences have sharpened my ability to effectively communicate with customers from a variety of backgrounds. I am a patient and empathetic listener, and I have been complimented on my aptitude for calmly negotiating positive solutions during confrontational situations.

Initiative

For the Valleyview Walk for Arthritis, I created and facilitated two new fundraising events: a cake auction and a pet show. Each event raised over $1000.00.

At McBride's Restaurant, I developed a simple computerized seating plan. I presented it to my manager, and she has used my plan in the restaurant for the past year.

At Sunset Seniors Home, I recently started an Agatha Christie book club that has been very successful. I was able to get donations of second-hand Agatha Christie books from a used bookstore.

Writing Skills

Throughout high school I maintained an A+ in my English classes. I am a composition tutor at Valleyview College. In order to be a tutor at Valleyview, students must maintain an A average. I have always enjoyed writing and have written many stories and articles.

Employment

| 2007 – present | McBride's Restaurant | **Cashier and Evening Manager** |
| 2005 – 2006 | Barney's Nearly New | **Sales Clerk and Cashier** |

Volunteer experience

2005 – present Sunset Seniors Centre   **Companion and Tea Server**

I visit the centre twice a week from 4 to 8 PM. I visit with residents and help serve evening tea.

2006 – present   Valleyview Walk for Arthritis   **Fundraiser, Server**

I participate in the fundraising prior to event day and serve refreshments at a pit stop on event day.

References

Available upon request

Spelling and Vocabulary

LEARNING OBJECTIVES:

After reading this chapter and completing the exercises in it, you will be able to

1. Use homonyms correctly.

2. Identify and correct common spelling errors.

3. Spell and define workplace words correctly.

4. Edit your work for errors not caught by the spell-check tool in your word processing software.

Introduction

By this stage of your educational career and life experience, you know whether or not you can spell well. People who read a lot are often good spellers, simply because they have been exposed to so much writing. People who like to write often spell well because they enjoy working with language.

In the community services field, you will complete a variety of writing tasks by hand: daily logs, meeting notes, off-site assessments or interviews (e.g., home visits), and court notes to name a few. This means that you will not always have the luxury of a spell checker. The handwritten notes you take will often be transcribed onto a computer, but for most situations, your handwritten notes are legal documents and must be kept as records. The odd spelling mistake can be excused; however, reports containing many spelling mistakes are unprofessional and undermine your credibility.

What can you do to ensure your writing is free of spelling errors? You must first acknowledge that spelling may be a weakness. Then, be proactive in producing error-free writing by taking the following steps:

1. Use a spell checker when you can.
2. Keep a dictionary at your desk.
3. When possible, have people proofread your work.
4. When possible, use vocabulary that you can spell confidently.
5. Learn to spell all the troublesome words identified in this chapter.

This chapter discusses three types of spelling errors and concludes with a list of workplace words you should learn to spell and define.

Commonly Confused Words

In the English language, many words sound the same but are spelled differently. These are called homonyms. There are also many words that sound enough alike to be commonly misused or misspelled. Here are several word sets with their correct meanings. (The homonyms are identified.)

accept/except

to <u>accept</u> means to agree with or be receptive to something: I *accept* your apology.
<u>except</u> means to excuse or omit something: Everyone *except* Paul needs the self-defence training update.

affect/effect

<u>affect</u> is a verb: We were greatly *affected* by her victim impact statement.
<u>effect</u> is a noun: His belligerent behaviour had a negative *effect* on his youth worker.
<u>effect</u> can also be a verb: Politicians effect change.

by/buy (homonyms)

<u>by</u> indicates place or ownership: The report was written *by* Sondra.
The stapler is *by* the coffee machine.
to <u>buy</u> is to purchase something: We need to *buy* more blankets for the shelter.

council/counsel (homonyms)

<u>council</u> is a noun: The advisory *council* voted to support the new life skills program.
<u>counsel</u> is a verb: The child and youth worker *counselled* her client to stay in school.

farther/further

<u>farther</u> describes distance: How much *farther* is it to the next off-ramp?
<u>further</u> describes the exploration of something in a more in-depth way: I think we should discuss Sarita's plan of care *further*.

principal/principle (homonyms)

a <u>principal</u> is a person in charge of an educational institution: The *principal* welcomed the new students with juice and muffins.

a <u>principal</u> is also used to modify a noun and means main or primary: The *principal* reason for his aggressive behaviour was his abusive upbringing.

a <u>principle</u> is an ethical or moral stance: I stand by my *principles* and will not accept your bribe.

stationary/stationery (homonyms)

<u>stationary</u> indicates lack of movement: The centre now has two *stationary* bikes to go with the other exercise equipment.

<u>stationery</u> describes writing materials: Zoe's mother brought her some new *stationery*, so she can write home.

their/there/they're (homonyms)

<u>their</u> is a possessive pronoun: *Their* parents will also attend the assessment interview.

<u>there</u> is an indication of place: You can park over *there*.

<u>they're</u> is the contraction of they are: *They're* meeting tomorrow.

to/too/two (homonyms)

<u>to</u> is an indication of place: He is moving *to* a new halfway house.

<u>too</u> means in addition to, also: She is taking the anger management course *too*.

<u>two</u> is a number: She has two sisters.

write/right/rite (homonyms)

to <u>write</u> is a verb: She must *write* the incident report before going off shift.

a <u>right</u> is a noun: It is your *right* to have a parole hearing.

to <u>right</u> is a verb: I would like to *right* all wrongs done to children.

a <u>rite</u> is a noun: Getting your driver's licence is a rite of passage.

Commonly Misspelled and Misused Words

Here are some commonly misspelled and misused words. Read through this list to see if you are guilty of making these errors.

Error	Correct Spelling
alright	all right
alot	a lot
could of (would of, should of)	could have

Canadian Spelling

Be sure to set the spell checker on your word processing program to Canadian English. There are several words that are spelled differently in Canada than they are in the United States. If your spell checker is not set to Canadian English, it will not catch these words and you will inadvertently make mistakes.

Here are some examples:

Canadian Spelling	US Spelling
centre	center
colour	color
cheque	check
counselling	counseling
labour	labor
litre	liter
licence (noun), license (verb)	license

Workplace Words

You should know the spelling and meaning of the following 200 workplace words.

List 1	List 2	List 3	List 4
Penitentiary	Lethargic	Anger	Delinquent
Subpoena	Allocate	Repeal	Alcohol
Recidivism	Alleviate	Testify	Suicide
Psychologist	Parole	Gerontology	Symptom
Psychiatrist	Addiction	Geriatric	Syndrome
Diagnosis	Hospice	Placement	Development
Supervise	Occurrence	Recommend	Dependence
Augment	Incident	Disability	Assimilation
Develop	Emergency	Occupational	Holistic
Licence	Prescription	Opportunity	Indigenous
Advocate	Medication	Accessibility	Spiritual
Abrasion	Dosage	Counsellor	Legislation
Laceration	Restraint	Transition	Linguistics
Initiate	Adolescent	Input	Denomination
Proceed	Assess	Personnel	Acculturation
Procedure	Access	Resource	Schizophrenia
Committee	Accuse	Summary	Psychopath
Commitment	Admission	Summarize	Sociopath
Concurrent	Administrate	Volunteer	Victimization
Reprieve	Attitude	Literacy	Harassment
Recognizance	Advise	Equipment	Acquaintance
Custody	Advice	Correspondence	Associates
Cooperate	Adhere	Reference	Intimate
Coordinator	Anticipate	Professional	Practitioner
Pastime	Aggravate	Aboriginal	Practical

List 5	List 6	List 7	List 8
Judgment	Anonymous	Commute	Testimony
Diabetic	Unruly	Liaise	Accumulate
Unkempt	Correctional	Precedent	Remission
Cognitive	Restorative	Statute	Recovery
Behavioural	Profession	Privilege	Beneficial
Reintegration	Investigation	Disclosure	Tolerance
Embarrassment	Judicial	Lawyer	Neighbour
Latitude	Solitary	Expertise	Custodial
Reassurance	Absence	Consequence	Extenuating
Claustrophobic	Deviance	Contentious	Culpable
Arraign	Socialization	Integrity	Program
Hysterical	Immigrant	Ethical	Conscientious
Relinquish	Grievance	Prejudice	Conscious
Argument	Mitigate	Allegation	Segregate
Facility	Consequences	Availability	Solitary
Disappointment	Interpersonal	Circumstance	Suspicious
Receipt	Maintenance	Responsibility	Qualitative
Potential	Discretion	Curriculum	Quantitative
Possessive	Ethnicity	Technician	Experiment
Compulsive	Prohibit	Outreach	Municipal
Inability	Character	Sponsor	Regional
Initiate	Sufficient	Relevant	Administer
Subsequent	Homicide	Reprobate	Convince
Impersonate	Supervision	Attest	Termination
Anonymity	Juvenile	Organize	Facilitate

Spell-Check Feature

The spell-check feature on your word processing program is a useful tool; however, as you know, a spell checker catches only spelling errors; it does not catch misused words.

ACTIVITY 9.1   Editing after the Spell Check

Consider the following paragraph:

> Last night their was a serious incident at the group home. Earlier in the day, Leila's principle had called too report her absence from school that day. When staff member Julie Cook spoke with Leila about it, Leila got very angry. Leila grabbed a heavy lamp from the office desk and could of seriously injured Julie if I hadn't of grabbed the lamp from Julie. Leila has acted out alot lately, and we have decided to meat with her caseworker to draft a revised plan of care.

There are seven spelling errors in this paragraph that a spell checker would not catch. Can you locate them?

_____

_____

_____

Your Personal Spelling Issues

We all have words that give us trouble when we try to spell them. Consider starting a list of these troublesome words and keeping it with you in a small journal. If you carry an agenda or address book, you can write the list there. Add to the list as needed. Cross words off the list when you have mastered them. This list will be handy in the workplace. You can add troublesome workplace words you often need to use.

Chapter Summary

- Professional writing requires careful attention to spelling.

- The community services field has many workplace-specific words that you will need to define and spell correctly.

- Several words are spelled differently in Canada than they are in the United States; always use Canadian spelling for your workplace writing.

- Take the time to carefully edit your work to ensure homonyms and words that sound alike are used and spelled correctly.

Chapter Exercises

EXERCISE 9.1 ## Choosing the Appropriate Word

Choose the correct word from the italicized word pairs in each of the following sentences.

1. She was emotionally *effected/affected* by her daughter's illness.

2. Please drop by the *centre/center* after you finish with your client.

3. *There/Their/They're* case is up next before the judge.

4. *By/Buy* the time we reached the client's home, she was calm and ready to speak to us.

5. I have been asked to *council/counsel* our new resident.

6. We have *alot/a lot* of work ahead of us tonight.

7. Roger has lost his *licence/license* because he was caught street-racing.

8. The *principal/principle* reason for my client's current aggressive behaviour is his previous abusive foster-home situation.

9. I have taken *to/too* many sick days because of my chronic asthma.

10. Our new client likes to *colour/color* on the walls when no one is looking.

EXERCISE 9.2 Defining Words Correctly

For each of the following words, write sentences that clearly show the word's meaning. The first one is completed as an example.

1. they're: *They're bringing our new office furniture today.*

   _____

   _____

2. affect:

   _____

   _____

3. council:

   _____

   _____

4. right:

   _____

   _____

5. principle:

   _____

   _____

6. further:

   _____

   _____

7. except:

   _____

   _____

8. buy:

   _____

   _____

9. stationary:

   _____

   _____

10. too:

   _____

   _____

EXERCISE 9.3 Expanding Your Vocabulary

Over a period of a few weeks, look up all the words from the workplace word list on pages 158–159. Make sure you understand the proper meaning for each word.

Look up all the words, even if you are sure you know the meaning of a word. Sometimes words are commonly misused; for example, people often use the word *delinquent* to describe a teenage person who is causing trouble or who is in conflict with the law. The dictionary definition of *delinquent* does not specifically reference teens or juveniles.

Write out the definitions of any words for which you did not know the meanings.

Take this task slowly. Spend only about 30 minutes at a time on this exercise to effectively absorb and retain the new definitions.

APPENDIX   Research Skills and Plagiarism

An important component of your college work will be the research you do for papers, projects, reports, and presentations. Once you are working in the community services field, you will also have occasion to do research. Although conducting research can be an interesting and enriching experience, for many, research assignments cause stress for two reasons:

1. Students are unsure how to find valid sources
2. Students are unsure how to properly credit sources

This appendix provides research tips to help you with your research projects. Before you read the tips, it's important to have a clear understanding of what constitutes plagiarism.

What is Plagiarism?

The definition of plagiarism is straightforward: taking the work of others and pretending it is your own. Plagiarism is stealing.

Most students understand that if they quote someone directly from a text (print or online), they must put quotation marks around the borrowed idea and add a citation that tells the reader where the information originally came from.

Where students get into trouble is when they put someone's idea into their own words. Many students feel that the act of changing the wording is sufficient to count that work as one's own. This is not so. If you use someone else's idea, whether as a direct quote or a paraphrase, you must tell your reader where that idea originally came from.

As well, using pictures, statistics, graphs, music, computer code, etc., from other sources without giving credit is also plagiarism.

Research Tips

1. When you are assigned a research topic or have chosen a research topic, **do no research until you've written down everything you know about your topic.** This way, you will have a solid base of your own work that you know isn't plagiarized.

2. If you are assigned a topic about which you know nothing, ask your teacher how she would like you to handle the research. Because everything you write will come from another source, will your teacher want you to give citations after every sentence? After every few sentences? After every paragraph? Ideally, teachers won't give assignments in which all you are doing is research. Good research assignments ask you to find out about a topic and then apply what you've learned to situations, case studies, or your opinion. This way your assignment has a good balance of your own material and research material.

3. Aim for a good variety of resource material: books, journals, newspapers, electronic sources, etc.

4. Choose your Internet sources wisely. Make sure your sources are credible. This means checking for the credentials of your source. Have you accessed a university site written by experts? Have you accessed a blog written by passionate amateurs? Both can give you legitimate material, but you must understand the value of each and illustrate that value in your assignment. It also means being aware of any bias in the resources you choose. Have you accessed a site that has a political or advertorial stance on a product or service? Have you accessed a site that presents a balanced, objective view? Again, there is no problem using both types of resources if you understand the biases and account for them in your assignment.

5. Wikipedia is an online encyclopedia. As with print encyclopedias, online encyclopedias are an excellent jumping off point for your research. Encyclopedias provide a summary of topics and provide leads for further research.

   The unique feature about Wikipedia is that it is available for anyone to add to. This means that you can go onto any Wikipedia page and edit (change) the information that is there. This is why many teachers will ask that you do not use Wikipedia as a resource. They are worried that false information will be on the topic pages you access. This can happen.

   However, among many scholars, Wikipedia has the reputation of being a reliable source. It is typically accessed and added to by experts who are passionate about sharing well-researched and up-to-date information. It is constantly monitored, and in the few instances where people have sabotaged pages with faulty information, regular contributors to those pages have restored the pages back to their original quality. One useful feature on Wikipedia is the History button. This is a link to the history of all contributions. On the History page readers can find out the names of contributors, their credentials, and the times material was added or edited.

   If you like using Wikipedia, ask your teacher if you can include it as a legitimate reference. If your teacher does not recognize Wikipedia as a legitimate source, you can still use it to inform yourself about you topic and get some ideas for further research.

6. Make use of your school's library or learning resource centre (LRC). The staff in your library or LRC have specialized training in research and are employed to help you with your research. They will be able to direct you to appropriate databases and print collections. As well, many college and university libraries or LRCs have borrowing agreements with other institutions, which means that you have access to countless materials. Make it a point to see what your library or LRC has to offer. Your school's library or LRC will also have a website with valuable links to research, writing, and documentation tutorials and resources such as Canadian newspapers, Statistics Canada, and federal and provincial ministry websites.

7. As you research, be sure to record all of the information you will need in order to properly reference your sources during the final draft of your assignment.
   - For printed material, write down titles and subtitles, authors' names, page numbers, volume numbers, edition numbers, and publisher's names and locations.
   - For electronic sources, write down the date of the website, the date you accessed your information, page or paragraph numbers, authors' names, titles and subtitles, and web addresses.

8. Find out from your teachers which style of documentation they want you to use. Typically, the community services field uses the APA style (American Psychological Association). Confirm this with your teachers.

9. Refer to a reliable book or Internet source when writing your in-text citations and reference page. Many universities and colleges have on-line information about how to set up your references.

10. Enjoy the research process! Think of it as a treasure hunt. Do not be shy about asking your teachers for help with finding valid sources or documenting your references. We like to be consulted!

INDEX